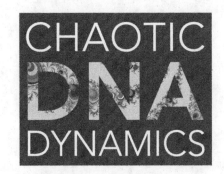

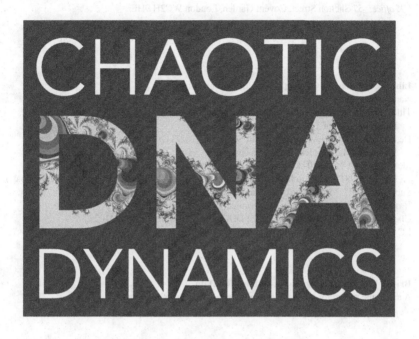

# Amujuri Mary Selvam

Indian Institute of Tropical Meteorology, India

**World Scientific**

NEW JERSEY · LONDON · SINGAPORE · BEIJING · SHANGHAI · HONG KONG · TAIPEI · CHENNAI · TOKYO

*Published by*

World Scientific Publishing Co. Pte. Ltd.

5 Toh Tuck Link, Singapore 596224

*USA office:* 27 Warren Street, Suite 401-402, Hackensack, NJ 07601

*UK office:* 57 Shelton Street, Covent Garden, London WC2H 9HE

**Library of Congress Cataloging-in-Publication Data**
Names: Selvam, A. M., author.
Title: Chaotic DNA dynamics / Amujuri Mary Selvam,
    Indian Institute of Tropical Meteorology, India.
Description: First. | New Jersey : World Scientific, [2022] |
    Includes bibliographical references and index.
Identifiers: LCCN 2021043719 | ISBN 9789811242854 (hardcover) |
    ISBN 9789811242861 (ebook for institutions) | ISBN 9789811243257 (ebook for individuals)
Subjects: LCSH: Meteorology--Research. | DNA--Structure. | System theory. |
    Chaotic behavior in systems. | Dynamical systems.
Classification: LCC QC869 .S45 2022 | DDC 003/.857--dc23/eng/20211102
LC record available at https://lccn.loc.gov/2021043719

**British Library Cataloguing-in-Publication Data**
A catalogue record for this book is available from the British Library.

For any available supplementary material, please visit
https://www.worldscientific.com/worldscibooks/10.1142/12463#t=suppl

Printed in Singapore

*To My Husband A. S. R. Murty*

# Preface

Current concepts in meteorological theory and limitations are as follows. The non-equilibrium system of atmospheric flows is modelled with assumption of local thermodynamic equilibrium up to the stratopause at 50 km; molecular motion of atmospheric component gases is implicitly embodied in the gas constant. Non-equilibrium systems can be studied numerically, but despite decades of research, it is still very difficult to define the analytical functions from which to compute their statistics and have an intuition for how these systems behave. Realistic mathematical modelling for simulation and prediction of atmospheric flows requires alternative theoretical concepts and analytical or error-free numerical computational techniques and therefore comes under the field of "General Systems Research" as explained in the following.

Space-time power law scaling and non-local connections exhibited by atmospheric flows have also been documented in other non-equilibrium dynamical systems, e.g., financial markets, neural network of brain, genetic networks, internet, road traffic, flocking behaviour of some animals and birds. Such universal behaviour has been subject of intensive study in recent years as "complex systems" under the subject headings self-organized criticality, nonlinear dynamics and chaos, network theory, pattern formation, information theory, cybernetics (communication, control and adaptation). Complex system is a system composed of many interacting parts, such that the collective behaviour or "emergent" behaviours of those parts together is more than the sum of their individual behaviours. Weather and climate are emergent properties of the complex adaptive system of atmospheric flows. Complex systems in different fields of study exhibit similar characteristics and therefore belong to the field of "General Systems". The terms "general systems" and "general systems research (or general systems theory)" are due to Ludwig von Bertalanffy. According to Bertalanffy, general systems research is a discipline whose subject matter is "the formulation and derivation of those principles which are valid for "systems' in general".

Skyttner (2006) quotes basic ideas of general systems theory formulated by Fredrich Hegel (1770–1831) as follows:

  (i)  The whole is more than the sum of the parts
 (ii)  The whole defines the nature of the parts
(iii)  The parts cannot be understood by studying the whole
(iv)  The parts are dynamically interrelated or interdependent

Skyttner, L. (2006) General Systems Theory: Problems, Perspectives, Practice (2nd Edition), World Scientific, 536pp.

In cybernetics, a system is maintained in dynamic equilibrium by means of communication and control between the constituent parts and also between the system and its environment.

This book is based on the author's research papers published in conference proceedings and research journals during her tenure (1962–1999) in the Indian Institute of Tropical Meterology, Pune 411008, India.

Chapter 1 deals with universal characteristics of fractal fluctuations as given by the general systems theory developed by the author. The spatial distribution of DNA base sequence A, C, G and T exhibit self-similar fractal fluctuations concomitant with inverse power law form for power spectra generic to dynamical systems in nature such as fluid flows, stock market fluctuations, population dynamics, etc. The physics of long-range correlations exhibited by fractals is not yet identified. A recently developed general systems theory visualizes the eddy continuum underlying fractals to result from the growth of large eddies as the integrated mean of enclosed small-scale eddies, thereby generating a hierarchy of eddy circulations, or an inter-connected network with associated long-range correlations. The model predictions are as follows: (i) The probability distribution and power spectrum of fractals follow the same inverse power law which is a function of the golden mean. The predicted inverse power law distribution is very close to the statistical normal distribution for fluctuations within two standard deviations from the mean of the distribution. (ii) Fractals signify quantum-like chaos since variance spectrum represents probability density distribution, a characteristic of quantum systems such as electron or photon. (ii) Fractal fluctuations of frequency distribution of DNA base sequence A, C, G and T signify spontaneous organization of the spatial distribution of DNA base sequence into the ordered pattern of the quasiperiodic Penrose tiling pattern. The model predictions are in agreement with the probability distributions and power spectra for the six different Genomic DNA data sets given in Chapters 3–8.

In Chapter 2, the topic of *Nonlinear dynamics, chaos and self-organized criticality* is discussed. Irregular (nonlinear) fluctuations on all scales of space and time are generic to dynamical systems in nature such as fluid flows, atmospheric weather patterns, heart beat patterns, stock market fluctuations, the DNA sequence of letters A, C, G and T, etc. Mandelbrot (1977) coined the name *fractal* for the non-Euclidean geometry of such fluctuations which have fractional dimension, for example, the rise and subsequent fall with time of the Dow Jones Index or rainfall that traces a zig-zag line in a two-dimensional plane and therefore has a *fractal* dimension greater than one but less than two. Mathematical models of dynamical systems are nonlinear and finite precision computer realizations exhibit sensitive dependence on initial conditions resulting in chaotic solutions, identified as deterministic chaos. *Nonlinear dynamics and chaos* is now (since 1980s) an area of intensive research in all branches of science. The *fractal* fluctuations exhibit scale invariance or self-similarity manifested as the widely documented inverse power law form for power spectra of space-time fluctuations identified as *self-organized criticality*. The power law is a distinctive experimental signature seen in a wide variety of complex systems. In economy it goes by the name fat tails, in physics it is referred to as critical fluctuations, in computer science and biology it is the edge of chaos, and in demographics it is called *Zipf*'s law. Power-law scaling is not new to economics. The power-law distribution of wealth discovered by *Vilfredo Pareto* (1848–1923) in the 19th predates any power laws in physics. One of the oldest scaling laws in geophysics is the *Omori law*. It describes the temporal distribution of the number of aftershocks, which occur after a larger earthquake (i.e., mainshock) by a scaling relationship. The other basic empirical seismological law, the *Gutenberg–Richter law* is also a scaling relationship, and relates intensity to its probability of occurrence. Time series analyses of global market economy also exhibits power-law behaviour with possible *multifractal* structure and has suggested an analogy to fluid turbulence. The observed power law represents structures similar to *Elliott waves* of technical analysis first introduced in the 1930s. It describes the time series of a stock price as made of different waves; these waves are in relation to each other through the *Fibonacci* series. *Elliott waves* could be a signature of an underlying critical structure of the stock market. Incidentally, the *Fibonacci* series represent a *fractal* tree-like branching network of self-similar structures. The commonly found shapes in nature are the helix and the dodecahedron, which are signatures of self-similarity underlying *Fibonacci* numbers. The general systems theory presented in this paper shows (Chapter 1) that *Fibonacci* series underlies *fractal* fluctuations on all space-time scales.

Chapter 3, Data I discusses universal spectrum for DNA base C+G frequency distribution in Human chromosomes 1–24. Power spectra of human DNA base C+G frequency distribution in all available contiguous sections exhibit the universal inverse power law form of the statistical normal distribution for the 24 chromosomes. Inverse power law form for power spectra of space-time fluctuations is generic to dynamical systems in nature and indicate long-range space-time correlations. The general systems theory (Chapter 1) predicts the observed non-local connections as intrinsic to quantumlike chaos governing space-time fluctuations of dynamical systems. The model predicts the following. (1) The quasiperiodic Penrose tiling pattern for the nested coiled structure of the DNA molecule in the chromosome resulting in maximum packing efficiency. (2) The DNA molecule functions as a unified whole fuzzy logic network with ordered two-way signal transmission between the coding and non-coding regions. Recent studies indicate influence of non-coding regions on functions of coding regions in the DNA molecule.

In Chapter 4, Data set II exhibits quantumlike chaos in the frequency distributions of bases A, C, G, T in human chromosome1 DNA. Recent studies of DNA sequence of letters A, C, G and T exhibit the inverse power law form frequency spectrum. Inverse power law form of the power spectra of fractal space-time fluctuations is generic to the dynamical systems in nature and is identified as self-organized criticality (Chapters 1 and 2). In this study it is shown that the power spectra of the frequency distributions of bases A, C, G, T in the Human chromosome 1 DNA exhibit self-organized criticality. DNA is a quasicrystal possessing maximum packing efficiency in a hierarchy of spirals or loops. Self-organized criticality implies that non-coding *introns* may not be redundant, but serve to organize the effective functioning of the coding *exons* in the DNA molecule as a complete unit.

In Chapter 5, Data set III exhibits universal spectrum for DNA base C+G concentration variability in Human chromosome Y. The spatial distribution of DNA base sequence A, C, G and T exhibit self-similar fractal fluctuations and the corresponding power spectra follow inverse power law of the form $1/f^{\alpha}$ where $f$ is the frequency and $\alpha$ the exponent. Inverse power law form for power spectra implies the following: (1) A scale invariant eddy continuum, namely, the amplitudes of component eddies are related to each other by a scale factor alone. In general, the scale factor is different for different scale ranges and indicates a multifractal structure for the spatial distribution of DNA base sequence. (2) Scale invariance of eddies also implies long-range spatial correlations of the eddy fluctuations. Multifractal structure to space-time fluctuations and the associated inverse power law form for power spectra is generic to spatially extended dynamical systems in nature and is a signature of self-organized criticality (Chapter 2). The author has developed a general systems theory (Chapter 1) where quantum mechanical laws emerge as self-consistent explanations for the observed long-range space-time correlations in macro-scale dynamical systems, i.e., the apparently chaotic fractal fluctuations are signatures of quantum-like chaos in dynamical systems. The model also provides unique quantification for the observed inverse power law form for power spectra in terms of the statistical normal distribution. In this paper it is shown that the frequency distribution of the bases C+G in all available contiguous sequences for Human chromosome Y DNA exhibit model predicted quantum-like chaos.

In Chapter 6, Data set IV quantumlike chaos is identified in the frequency distributions of the bases A, C, G, T in Drosophila DNA. Continuous periodogram power spectral analyses of *fractal* fluctuations of frequency distributions of bases A, C, G, T in Drosophila DNA show that the power spectra follow the universal inverse power-law form of the statistical normal distribution. Inverse power law form for power spectra of space-time fluctuations is generic to dynamical systems in nature and is identified as *self-organized criticality* (Chapter 2). The author has developed a general systems theory (Chapter 1), which provides universal quantification for observed *self-organized criticality* in terms of the statistical normal distribution. The long-range correlations intrinsic to *self-organized criticality* in macro-scale dynamical systems are a signature of quantumlike chaos. The *fractal* fluctuations self-organize to form an overall logarithmic spiral trajectory with the quasiperiodic *Penrose* tiling pattern for the internal structure. Power spectral analysis resolves such a spiral trajectory as an eddy continuum with embedded dominant wavebands. The dominant peak periodicities are functions of the *golden mean*. The observed *fractal* frequency distributions of the Drosophila DNA base sequences exhibit quasicrystalline structure with long-range spatial correlations or *self-organized criticality*. Modification of the DNA base sequence structure at any location may have significant noticeable effects on the function of the DNA molecule as a whole. The presence of noncoding *introns* may not be redundant, but serve to organize the effective functioning of the coding *exons* in the DNA molecule as a complete unit.

In Chapter 7, Data set V universal spectrum for DNA base CG frequency distribution in *Takifugu Rubripes* (Puffer fish) Genome is identified. The frequency distribution of DNA bases A, C, G, T exhibit fractal fluctuations ubiquitous to dynamical systems in nature. The power spectra of fractal fluctuations exhibit inverse power law form signifying long-range correlations between local (small-scale) and global (large-scale) perturbations. The author has developed a general systems theory (Chapter 1) model based on classical statistical physics for fractal fluctuations which predicts that the probability distribution of eddy amplitudes and the variance (square of eddy amplitude) spectrum of fractal fluctuations follow the universal Boltzmann inverse power law (for molecular energies) expressed as a function of the golden mean. The model predicted distribution is very close to statistical normal distribution for fluctuations within two standard deviations from the mean and exhibits a fat long tail. In this chapter, it is shown that DNA base CG frequency distribution in *Takifugu rubripes* (Puffer fish) Genome Release 4 exhibit universal inverse power law form consistent with model prediction. The observed long-range correlations in the DNA bases implies that the non-coding "junk" or "selfish" DNA which appear to be redundant, may also contribute to the efficient functioning of the protein coding DNA, a result supported by recent studies.

In Chapter 8, Data set 6 long-range correlations are identified in Human chromosome X DNA base CG frequency distribution. Recent studies of DNA sequence of letters A, C, G and T exhibit the inverse power law form frequency spectrum. Inverse power law form of the power spectra of fractal space-time fluctuations is generic to the dynamical systems in nature and is identified as self-organized criticality (Chapter 2). In this study it is shown that the power spectra of the frequency distributions of bases C+G in the Human chromosome X DNA exhibit self-organized criticality. DNA is a quasicrystal possessing maximum packing efficiency in a hierarchy of spirals or loops. Self-organized criticality implies that non-coding *introns* may not be redundant, but serve to organize the effective functioning of the coding *exons* in the DNA molecule as a complete unit.

There is close agreement between the model predicted and the observed universal spectrum for the fractal fluctuations of the different sets of Genomic DNA base sequences. Universal spectrum for fractal fluctuations of DNA base sequence implies that DNA molecule functions as a unified whole communicating network with long-range correlations between the component base sequences for the effective functioning of the living system as a whole. The so-called "junk DNA" which appear to be redundant with no specific role also are a vital part of the fractal network for communication and control of essential tasks.

A. M. Selvam
*Deputy Director (Retired)*
*Indian Institute of Tropical Meteorology*
*Pune 411008, India*
amselvam@gmail.com

# Contents

# Chapter 1

# Universal Characteristics of Fractal Fluctuations: General Systems Theory

## 1.1 Introduction

Dynamical systems in nature such as atmospheric flows, heartbeat patterns, population dynamics, stock market indices, DNA base A, C, G, T sequence pattern, prime number distribution, etc., exhibit irregular (chaotic) space-time fluctuations on all scales and exact quantification of the fluctuation pattern for predictability purposes has not yet been achieved. The irregular fluctuations, however manifest a new kind of order, that of self-similarity, i.e., the larger-scale fluctuations resemble in shape the enclosed smaller-scale fluctuations signifying long-range correlations seen as inverse power-law form for power spectra of the fluctuations. The fractal or self-similar nature of space-time fluctuations was identified by Mandelbrot (1975) in the 1970s. Representative examples of fractal fluctuations of (i) daily percentage change of Dow Jones Index (ii) human DNA base CG concentration/10 bp (base pairs) (iii) *Takifugu rubripes* (Puffer fish) DNA base CG concentration/10 bp are shown in Fig. 1.1.

Fractal fluctuations (Fig. 1.1) show a zigzag self-similar pattern of successive increase followed by decrease on all scales (space-time), for example, in atmospheric flows, cycles of increase and decrease in meteorological parameters such as wind, temperature, etc., occur from the turbulence scale of millimeters-seconds to climate scales of thousands of

1

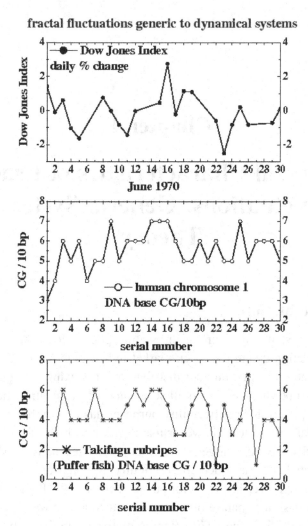

**Figure 1.1:** Representative examples of fractal fluctuations of (i) daily percentage change of Dow Jones Index (ii) human DNA base CG concentration/10 bp (iii) *Takifugu rubripes* (Puffer fish) DNA base CG concentration/10 bp.

kilometres-years. The power spectra of fractal fluctuations exhibit $1/f$ noise (Planat, 2003) manifested as inverse power-law of the form $f^{-\alpha}$ where $f$ is the frequency and $\alpha$ is a constant. Inverse power-law for power spectra indicate long-range space-time correlations or scale invariance for the scale range for which $\alpha$ is a constant, i.e., the amplitudes of the eddy fluctuations in this scale range are a function of the scale factor $\alpha$ alone. In general, the value of $\alpha$ is different for different scale ranges indicating multifractal structure for the fluctuations. The long-range space-time correlations exhibited by dynamical systems are identified as self-organized criticality (Bak *et al.*, 1988; Schroeder, 1990). $1/f$ fluctuations occur in areas as diverse as electronics, chemistry, biology, cognition or geology and claims for an unifying mathematical principle (Milotti, 2002; Planat *et al.*, 2002; Planat, 2003). The physics of fractal fluctuations generic to dynamical systems in nature is not yet identified and traditional statistical, mathematical theories do not provide adequate tools for identification and quantitative description of the observed universal properties of fractal structures observed in all fields of science and other areas of human interest. A recently developed general systems theory for fractal space-time fluctuations (Selvam, 1990, 2005, 2007; Selvam and Fadnavis, 1998; Selvam, 2009, 2011) shows that the larger-scale fluctuation can be visualized to emerge from the space-time averaging of enclosed small-scale fluctuations, thereby generating a hierarchy of self-similar fluctuations manifested as the observed eddy continuum in power spectral analyses of fractal fluctuations. Such a concept results in inverse power-law form incorporating the *golden mean* $\tau$ for the space-time fluctuation pattern and also for the power spectra of the fluctuations (Section 1.4). The predicted distribution is close to the Gaussian distribution for small-scale fluctuations, but exhibits *fat long tail* for large-scale fluctuations. The general systems theory, originally developed for turbulent fluid flows, provides universal quantification of physics underlying fractal fluctuations and is applicable to all dynamical systems in nature independent of its physical, chemical, electrical, or any other intrinsic characteristic. In the following, Section 1.2 gives a summary of traditional statistical and mathematical theories/techniques used for analysis and quantification of space-time fluctuation data sets. The drawbacks of existing techniques of data quantification are discussed and important model predictions of the general systems theory are listed. The general systems theory for fractal space-time fluctuations originally developed for turbulent fluid flows is described in Section 1.3. The universal Feigenbaum's constants $a$ and $d$ characterizing dynamical systems is incorporated in model predictions for probability density distribution function for fractal fluctuations in

Section 1.4. The applications of general systems theory concepts to spatial distribution of different Genomic DNA base sequences are discussed in Chapters 3–8.

## 1.2  Statistical Methods for Data Analysis

Dynamical systems such as atmospheric flows, stock markets, heartbeat patterns, population growth, traffic flows, etc., exhibit irregular space-time fluctuation patterns. Quantification of the space-time fluctuation pattern will help predictability studies, in particular for events which affect day-to-day human life such as extreme weather events, stock market crashes, traffic jams, etc. The analysis of data sets and broad quantification in terms of probabilities belongs to the field of statistics. Early attempts resulted in identification of the following two quantitative (mathematical) distributions which approximately fit data sets from a wide range of scientific and other disciplines of study. The first is the well-known statistical normal distribution and the second is the power-law distribution associated with the recently identified *fractals* or self-similar characteristic of data sets in general. Traditionally, the Gaussian probability distribution is used for a broad quantification of the data set variability in terms of the sample mean and variance. In the following, a summary is given of the history and merits of the two distributions.

### 1.2.1  *Statistical normal distribution*

Selvam (2009, 2011) has discussed the applicability of statistical normal distribution to fractal data sets as follows. Historically, our present day methods of handling experimental data have their roots about 400 years ago. At that time scientists began to calculate the odds in gambling games. From those studies emerged the theory of probability and subsequently the theory of statistics. These new statistical ideas suggested a different and more powerful experimental approach. The basic idea was that in some experiments random errors would make the value measured a bit higher and in other experiments random errors would make the value measured a bit lower. Combining these values by computing the average of the different experimental results would make the errors cancel and the average would be closer to the "right" value than the result of any one experiment (Liebovitch and Scheurle, 2000).

Abraham de Moivre, an 18th century statistician and consultant to gamblers made the first recorded discovery of the normal curve of error (or the bell curve because of its shape) in 1733. The normal distribution is the limiting case of the binomial distribution resulting from random operations such as flipping coins or rolling dice. Serious interest in the distribution of errors on the part of mathematicians such as Laplace and Gauss awaited the early 19th century when astronomers found the bell curve to be a useful tool to take into consideration the errors they made in their observations of the orbits of the planets (Goertzel and Fashing, 1981). The importance of the normal curve stems primarily from the fact that the distributions of many natural phenomena are at least approximately normally distributed. This normal distribution concept has moulded how we analyze experimental data over the last 200 years. We have come to think of data as having values most of which are near an average value, with a few values that are smaller, and a few that are larger. The probability density function PDF($x$) is the probability that any measurement has a value between $x$ and $x + dx$. We suppose that the PDF of the data has a normal distribution. Most quantitative research involves the use of statistical methods presuming *independence* among data points and Gaussian "normal" distributions (Andriani and McKelvey, 2007). The Gaussian distribution is reliably characterized by its stable mean and finite variance (Greene, 2002). Normal distributions place a trivial amount of probability far from the mean and hence the mean is representative of most observations. Even the largest deviations, which are exceptionally rare, are still only about a factor of two from the mean in either direction and are well characterized by quoting a simple standard deviation (Clauset *et al.*, 2007). However, apparently rare real-life catastrophic events such as major earth quakes, stock market crashes, heavy rainfall events, etc., occur more frequently than indicated by the normal curve, i.e., they exhibit a probability distribution with a *fat tail*. Fat tails indicate a power-law pattern and interdependence. The "tails" of a power-law curve — the regions to either side that correspond to large fluctuations — fall off very slowly in comparison with those of the bell curve (Buchanan, 2004). The normal distribution is therefore an inadequate model for extreme departures from the mean.

The following references are cited by Goertzel and Fashing (1981) to show that the bell curve is an empirical model without supporting theoretical basis: (i) Modern texts usually recognize that there is no theoretical justification for the use of the normal curve, but justify using it as a convenience (Cronbach, 1970) (ii) The bell curve came to be generally accepted, as M. Lippman remarked to Poincare (1968), because "... the experimenters fancy that it is a theorem in mathematics and the mathematicians that it is an experimental fact" (iii) Karl Pearson (best known today for the invention of the product-moment correlation coefficient) used his newly developed *Chi Square* test to check how closely a number of empirical distributions of supposedly random errors fitted the bell curve. He found that many of the distributions that had been cited in the literature as fitting the normal curve were actually significantly different from it, and concluded that "the normal curve of error possesses no special fitness for describing errors or deviations such as arise either in observing practice or in nature" (Pearson, 1900).

## 1.2.2 *Fractal fluctuations and statistical analysis*

Fractals are the latest development in statistics. The space-time fluctuation pattern in dynamical systems was shown to have a self-similar or fractal structure in the 1970s (Mandelbrot, 1975, 1977). The larger-scale fluctuation consists of smaller-scale fluctuations identical in shape to the larger scale. An appreciation of the properties of fractals is changing the most basic ways we analyze and interpret data from experiments and is leading to new insights into understanding physical, chemical, biological, psychological, and social systems. Fractal systems extend over many scales and so cannot be characterized by a single characteristic average number (Liebovitch and Scheurle, 2000). Further, the self-similar fluctuations imply long-range space-time correlations or interdependence. Therefore, the Gaussian distribution will not be applicable for description of fractal data sets. However, the bell curve still continues to be used for approximate quantitative characterization of data which are now identified as fractal space-time fluctuations.

### 1.2.2.1 *Power-laws and fat tails*

Fractals conform to power-laws. A power-law is a relationship in which one quantity $A$ is proportional to another $B$ taken to some power $n$; that is, $A \sim B^n$ (Buchanan, 2004). One of the oldest scaling laws in geophysics is the Omori law (Omori, 1895). This law describes the temporal distribution of the number of after-shocks, which occur after a larger earthquake (i.e., main-shock) by a scaling relationship. Richardson (1960) came close to the concept of fractals when he noted that the estimated length of an irregular coastline scales with the length of the measuring unit. Andriani and McKelvey (2007) have given exhaustive references to earliest known work on power-law relationships summarized as follows. Pareto (1897) first noticed power-laws and fat tails in economics. Cities follow a power-law when ranked by population (Auerbach, 1913). Dynamics of earth-quakes follow power-law (Gutenberg and Richter, 1944) and Zipf (1949) found that a power-law applies to word frequencies (Estoup, 1916) had earlier found a similar relationship). Mandelbrot (1963) rediscovered them in the 20th century, spurring a small wave of interest in finance (Fama, 1965; Montroll and Shlesinger, 1984). However, the rise of the "standard" model (Gaussian) of efficient markets, sent power-law models into obscurity. This lasted until the 1990s, when the occurrence of cata-strophic events, such as the 1987 and 1998 financial crashes, that were difficult to explain with the "standard" models (Bouchaud *et al.*, 1998), re-kindled the fractal model (Mandelbrot and Hudson, 2004).

There are many physical and/or mathematical mechanisms that gener-ate power-law distributions and self-similar behavior. Understanding how a mechanism is selected by the microscopic laws constitutes an active field of research (Sornette, 2007). Sornette *et al.* (1995) cite the works of Mandelbrot (1983), Aharony and Feder (1989) and Riste and Sherrington (1991) and state that observation that many natural phenomena have size distributions that are power-laws, has been taken as a fundamental indica-tion of an underlying self-similarity. A power-law distribution indicates the absence of a characteristic size and as a consequence that there is no upper limit on the size of events. The largest events of a power-law distri-bution completely dominate the underlying physical process; for instance, fluid-driven erosion is dominated by the largest floods and most deformation at plate boundaries takes place through the agency of the largest earthquakes. It is a matter of debate whether power-law distributions, which are valid descriptions of the numerous small and intermediate events, can be extrapolated to large events; the largest events are, almost by definition, under-sampled.

A power-law world is dominated by extreme events ignored in a Gaussian world. In fact, the fat tails of power-law distributions make large extreme events orders-of-magnitude more likely. Theories explaining power-laws are also scale-free. This is to say, the same explanation (theory) applies at all levels of analysis (Andriani and McKelvey, 2007).

### 1.2.2.2  *Scale-free theory for power-laws with fat, long tails*

A scale-free theory for the observed fractal fluctuations in atmospheric flows shows that the observed long-range spatiotemporal correlations are intrinsic to quantum-like chaos governing fluid flows. The model concepts are independent of the exact details such as the chemical, physical, physiological and other properties of the dynamical system and therefore provide a general systems theory applicable to all real world and computed model dynamical systems in nature (Selvam, 1993; Selvam and Fadnavis, 1999a, 1999b; Selvam, 2001a, 2001b, 2002, 2004). The model is based on the concept that the irregular fractal fluctuations may be visualized to result from the superimposition of an eddy continuum, i.e., a hierarchy of eddy circulations generated at each level by the space-time integration of enclosed small-scale eddy fluctuations. Such a concept of space-time fluctuation averaged distributions *should* follow statistical normal distribution according to *Central Limit Theorem* in traditional statistical theory (Ruhla, 1992). Also, traditional statistical/mathematical theory predicts that the Gaussian, its Fourier transform and therefore Fourier transform associated power spectrum are the same distributions. The Fourier transform of normal distribution is essentially a normal distribution. A power spectrum is based on the Fourier transform, which expresses the relationship between time (space) domain and frequency domain description of any physical process (Phillips, 2005; Riley *et al.*, 2006). However, the general systems theory model (Section 1.3) visualizes the eddy growth process in successive stages of unit length-step growth with ordered two-way energy feedback between the larger and smaller-scale eddies and derives a power-law probability distribution $P$ which is close to the Gaussian for small deviations and gives the observed fat, long tail for large fluctuations. Further, the model predicts the power spectrum of the eddy continuum also to follow the power-law probability distribution $P$.

In summary, the model predicts the following:

(1) The eddy continuum consists of an overall logarithmic spiral trajectory with the quasiperiodic *Penrose* tiling pattern for the internal structure.

(2) The successively larger eddy space-time scales follow the Fibonacci number series.

(3) The probability distribution $P$ of fractal domains for the $n$th step of eddy growth is equal to $\tau^{-4n}$ where $\tau$ is the golden mean equal to $(1 + \sqrt{5})/2$ ($\approx 1.618$). The eddy growth step $n$ represents the *normalized deviation t* in traditional statistical theory. The *normalized deviation t* represents the departure of the variable from the mean in terms of the standard deviation of the distribution assumed to follow *normal distribution* characteristics for many real world space-time events. There is progressive decrease in the probability of occurrence of events with increase in corresponding *normalized deviation t*. Space-time events with *normalized deviation t* greater than 2 occur with a probability of 5% or less and may be categorized as extreme events associated in general with widespread (space-time) damage and loss. The model predicted probability distribution $P$ is close to the statistical normal distribution for $t$ values less than 2 and greater than normal distribution for $t$ more than 2, thereby giving a *fat, long tail*. There is non-zero probability of occurrence of very large events.

(4) The inverse of probability distribution $P$, namely, $\tau^{4n}$ represents the relative eddy energy flux in the large eddy fractal (small-scale fine structure) domain. There is progressive decrease in the probability of occurrence of successive stages of eddy growth associated with progressively larger domains of fractal (small-scale fine structure) eddy energy flux and at sufficiently large growth stage trigger catastrophic extreme events such as heavy rainfall, stock market crashes, traffic jams, etc., in real world situations.

(5) The power spectrum of fractal fluctuations also follows the same distribution $P$ as for the distribution of fractal fluctuations. The square of the eddy amplitude (variance) represents the eddy energy and therefore the eddy probability density $P$. Such a result that the additive amplitudes of eddies when squared represent probabilities, is exhibited by the sub-atomic dynamics of quantum systems such as the electron or proton (Maddox, 1988, 1993; Rae, 1988). The phase spectrum is the same as the variance spectrum, a characteristic of quantum systems identified as *Berry's phase*. Fractal fluctuations are signatures of quantum-like chaos in dynamical systems.

(6) The *fine structure constant* for spectrum of fractal fluctuations is a function of the *golden mean* and is analogous to that of atomic spectra and is equal to about 1/137.

(7) The universal algorithm for self-organized criticality is expressed in terms of the universal *Feigenbaum's* constants (Feigenbaum, 1980) $a$ and $d$ as $2a^2 = \pi d$ where the fractional volume intermittency of occurrence $\pi d$ contributes to the total variance $2a^2$ of fractal structures. The *Feigenbaum's* constants are expressed as functions of the *golden mean*. The probability distribution $P$ of fractal domains is also expressed in terms of the Feigenbaum's constants $a$ and $d$. The details of the model are summarized in the Sections 1.3 and 1.4 in the following.

## 1.3 General Systems Theory for Fractal Fluctuations

The fractal space-time fluctuations of dynamical systems may be visualized to result from the superimposition of an ensemble of eddies (sine waves), namely an eddy continuum. The relationship between large and small eddy circulation parameters are obtained on the basis of Townsend's (Townsend, 1956) concept that large eddies are envelopes enclosing turbulent eddy (small scale) fluctuations (Fig. 1.2).

The relationship between root mean square (r.m.s.) circulation speeds $W$ and $w_*$, respectively, of large and turbulent eddies of respective radii $R$ and $r$ is then given as

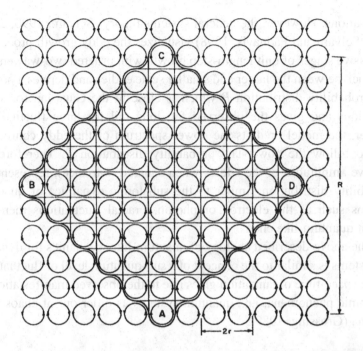

**Figure 1.2:**   Visualization of the formation of large eddy (ABCD) as envelope enclosing smaller scale eddies. By analogy, the continuum number field domain (Cartesian co-ordinates) may also be obtained from successive integration of enclosed finite number field domains.

$$W^2 = \frac{2}{\pi}\frac{r}{R}w_*^2 \qquad (1.1)$$

The dynamical evolution of space-time fractal structures is quantified in terms of ordered energy flow between fluctuations of all scales in Eq. (1.1), because the square of the eddy circulation speed represents the eddy energy (kinetic). A hierarchical continuum of eddies is generated by the integration of successively larger enclosed turbulent eddy circulations. Such a concept of space-time fluctuation averaged distributions *should* follow statistical normal distribution according to *Central Limit Theorem* in traditional Statistical theory (Ruhla, 1992). Also, traditional statistical/ mathematical theory predicts that the Gaussian, its Fourier transform and therefore Fourier transform associated power spectrum are the same distributions. However, the general systems theory (Selvam, 1990, 2005, 2007; Selvam and Fadnavis, 1998) visualizes the eddy growth process in successive stages of unit length-step growth with ordered two-way energy feedback between the larger and smaller scale eddies and derives a power-law probability distribution $P$ which is close to the Gaussian for small deviations and gives the observed fat, long tail for large fluctuations. Further, the model predicts the power spectrum of the eddy continuum also to follow the power-law probability distribution $P$. Therefore the additive amplitudes of the eddies when squared (variance), represent the probability distribution similar to the subatomic dynamics of quantum systems such as the electron or photon. Fractal fluctuations therefore exhibit quantum-like chaos.

The above-described analogy of quantum-like mechanics for dynamical systems is similar to the concept of a sub-quantum level of fluctuations whose space-time organization gives rise to the observed manifestation of subatomic phenomena, i.e., quantum systems as order out of chaos phenomena (Grossing, 1989).

### 1.3.1 *Dynamic memory (information) circulation network*

The r.m.s. circulation speeds $W$ and $w_*$ of large and turbulent eddies of respective radii $R$ and $r$ are related as

$$W^2 = \frac{2}{\pi}\frac{r}{R}w_*^2 \tag{1.1}$$

Equation (1.1) is a statement of the law of conservation of energy for eddy growth in fluid flows and implies a two-way ordered energy flow between the larger and smaller scales. Microscopic scale perturbations are carried permanently as internal circulations of progressively larger eddies. Fluid flows therefore act as dynamic memory circulation networks with intrinsic long-term memory of short-term fluctuations. Such *memory of water* is reported by Davenas *et al.* (1988).

## 1.3.2 *Quasicrystalline structure of the eddy continuum*

The turbulent eddy circulation speed and radius increase with the progressive growth of the large eddy (Selvam, 1990, 2005, 2007; Selvam and Fadnavis, 1998). The successively larger turbulent fluctuations, which form the internal structure of the growing large eddy, may be computed from Eq. (1.1) as

$$w_*^2 = \frac{\pi}{2} \frac{R}{dR} W^2 \qquad (1.2)$$

During each length step growth $dR$, the small-scale energizing perturbation $W_n$ at the $n$th instant generates the large-scale perturbation $W_{n+1}$ of radius $R$ where $R = \Sigma^n dR$ since successive length-scale doubling gives rise to $R$. Equation (1.2) may be written in terms of the successive turbulent circulation speeds $W_n$ and $W_{n+1}$ as

$$W_{n+1}^2 = \frac{\pi}{2} \frac{R}{dR} W_n^2 \qquad (1.3)$$

The angular turning $d\theta$ inherent to eddy circulation for each length step growth is equal to $dR/R$. The perturbation (spatial) $dR$ is generated by the small-scale acceleration $W_n$ (spatial displacement per unit time) at any instant $n$ and therefore $dR = W_n$. Starting with the unit value for $dR$ the successive $W_n$, $W_{n+1}$, $R$, and $d\theta$ values are computed from Eq. (1.3) and are given in Table 1.1.

**Table 1.1:** The computed spatial growth of the strange-attractor design traced by the macroscale dynamical system of atmospheric flows as shown in Fig. 1.3.

| R | $W_n$ | dR | d$\theta$ | $W_{n+1}$ | $\theta$ |
|---|---|---|---|---|---|
| 1.000 | 1.000 | 1.000 | 1.000 | 1.254 | 1.000 |
| 2.000 | 1.254 | 1.254 | 0.627 | 1.985 | 1.627 |
| 3.254 | 1.985 | 1.985 | 0.610 | 3.186 | 2.237 |
| 5.239 | 3.186 | 3.186 | 0.608 | 5.121 | 2.845 |
| 8.425 | 5.121 | 5.121 | 0.608 | 8.234 | 3.453 |

*(Continued)*

**Table 1.1.**   (*Continued*)

| R | $W_n$ | dR | d$\theta$ | $W_{n+1}$ | $\theta$ |
|---|---|---|---|---|---|
| 13.546 | 8.234 | 8.234 | 0.608 | 13.239 | 4.061 |
| 21.780 | 13.239 | 13.239 | 0.608 | 21.286 | 4.669 |
| 35.019 | 21.286 | 21.286 | 0.608 | 34.225 | 5.277 |
| 56.305 | 34.225 | 34.225 | 0.608 | 55.029 | 5.885 |
| 90.530 | 55.029 | 55.029 | 0.608 | 88.479 | 6.493 |

It is seen that the successive values of the circulation speed $W$ and radius $R$ of the growing turbulent eddy follow the Fibonacci mathematical number series such that $R_{n+1} = R_n + R_{n-1}$ and $R_{n+1}/R_n$ is equal to the golden mean $\tau$, which is equal to $((1 + \sqrt{5})/2) \cong 1.618$. Further, the successive $W$ and $R$ values form the geometrical progression $R_0(1 + \tau + \tau^2 + \tau^3 + \tau^4 + ....)$ where $R_0$ is the initial value of the turbulent eddy radius.

Turbulent eddy growth from primary perturbation $OR_O$ starting from the origin O (Fig. 1.3) gives rise to compensating return circulations $OR_1R_2$ on either side of $OR_O$, thereby generating the large eddy radius $OR_1$ such that $OR_1/OR_O = \tau$ and $R_OOR_1 = \pi/5 = R_OR_1O$. Therefore, short-range circulation balance requirements generate successively larger circulation patterns with precise geometry that is governed by the *Fibonacci* mathematical number series, which is identified as a signature of the universal period doubling route to chaos in fluid flows, in particular atmospheric flows. It is seen from Fig. 1.3 that five such successive length step growths give successively increasing radii $OR_1$, $OR_2$, $OR_3$, $OR_4$ and $OR_5$ tracing out one complete vortex-roll circulation such that the scale ratio $OR_5/OR_O$ is equal to $\tau^5 = 11.1$. The envelope $R_1R_2R_3R_4R_5$ (Fig. 1.3) of a dominant large eddy (or vortex roll) is found to fit the logarithmic spiral $R = R_0e^{b\theta}$ where $R_0 = OR_O$, $b = \tan \delta$ with $\delta$ the crossing angle equal to $\pi/5$, and the angular turning $\theta$ for each length step growth is equal to $\pi/5$. The successively larger eddy radii may be subdivided again in the *golden mean* ratio. The internal structure of large-eddy circulations is therefore made up of balanced small-scale circulations tracing out the well-known quasiperiodic *Penrose* tiling pattern identified as the quasicrystalline structure in condensed matter physics. A complete description of the atmospheric flow field is given by the quasiperiodic cycles with *Fibonacci* winding numbers.

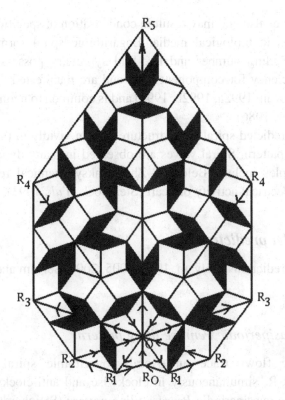

**Figure 1.3:**   The quasiperiodic *Penrose* tiling pattern with *five-fold* symmetry traced by the small eddy circulations internal to dominant large eddy circulation in turbulent fluid flows.

The quasiperiodic *Penrose* tiling pattern with five-fold symmetry has been identified as quasicrystalline structure in condensed matter physics (Janssen, 1988). The self-organized large eddy growth dynamics, therefore, spontaneously generates an internal structure with the five-fold symmetry of the dodecahedron, which is referred to as the *icosahedral* symmetry, e.g., the geodesic dome devised by *Buckminster Fuller*. Incidentally, the pentagonal dodecahedron is, after the helix, nature's second favourite structure (Stevens, 1974). Recently the carbon macromolecule $C_{60}$, formed by condensation from a carbon vapour jet, was found to exhibit the *icosahedral* symmetry of the closed soccer ball and has been named *Buckminsterfullerene* or *footballene* (Curl and Smalley, 1991). Self-organized quasicrystalline pattern formation therefore exists at the molecular level also and may result in condensation of specific biochemical structures in biological media. Logarithmic spiral formation with Fibonacci winding number and five-fold symmetry possess maximum packing efficiency for component parts and are manifested strikingly in *Phyllotaxis* (Jean, 1992a, 1992b, 1994) and is common to nature (Stevens, 1974; Tarasov, 1986).

Model predicted spiral flow structure is seen vividly in the hurricane cloud cover pattern. Spiral waves are observed in many dynamical systems. Examples include Belousov–Zhabotinksy chemical reaction and also in the electrical activity of heart (Steinbock *et al.*, 1993).

### 1.3.3 *Model predictions*

The model predictions (Selvam, 1990, 2005, 2007; Selvam and Fadnavis, 1998) are:

#### 1.3.3.1 *Quasiperiodic Penrose tiling pattern*

Atmospheric flows trace an overall logarithmic spiral trajectory $OR_0R_1R_2R_3R_4R_5$ simultaneously in clockwise and anti-clockwise directions with the quasiperiodic *Penrose tiling pattern* (Steinhardt, 1997) for the internal structure shown in Fig. 1.3.

The spiral flow structure can be visualized as an eddy continuum generated by successive length step growths $OR_0$, $OR_1$, $OR_2$, $OR_3$, ... respectively equal to $R_1$, $R_2$, $R_3$, ... which follow *Fibonacci* mathematical series such that $R_{n+1} = R_n + R_{n-1}$ and $R_{n+1}/R_n = \tau$ where $\tau$ is the *golden mean* equal to $(1 + \sqrt{5})/2$ ($\cong 1.618$). Considering a normalized length step equal to 1 for the last stage of eddy growth, the successively decreasing radial length steps can be expressed as 1, $1/\tau$, $1/\tau^2$, $1/\tau^3$, .... The normalized eddy continuum comprises of fluctuation length scales 1, $1/\tau$, $1/\tau^2$, .... The probability of occurrence is equal to $1/\tau$ and $1/\tau^2$, respectively, for eddy length scale $1/\tau$ in any one or both rotational (clockwise and anti-clockwise) directions. Eddy fluctuation length of amplitude $1/\tau$ has a probability of occurrence equal to $1/\tau^2$ in both rotational directions, i.e., the square of eddy amplitude represents the probability of occurrence in the eddy continuum. Similar result is observed in the subatomic dynamics of quantum systems which are visualized to consist of the superimposition of eddy fluctuations in wave trains (eddy continuum).

### 1.3.3.2 *Eddy continuum*

Conventional continuous periodogram power spectral analyses of such spiral trajectories in Fig. 1.3 ($R_oR_1R_2R_3R_4R_5$) will reveal a continuum of periodicities with progressive increase $d\theta$ in phase angle $\theta$ (theta) as shown in Fig. 1.4.

### 1.3.3.3 *Dominant eddies*

The broadband power spectrum will have embedded dominant wavebands ($R_oOR_1$, $R_1OR_2$, $R_2OR_3$, $R_3OR_4$, $R_4OR_5$, etc.) the bandwidth increasing

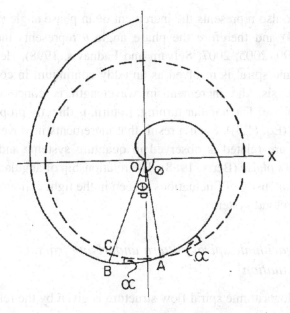

**Figure 1.4:** The equiangular logarithmic spiral given by $(R/r) = e^{\alpha\theta}$ where $\alpha$ and $\theta$ are each equal to $1/z$ for each length step growth. The eddy length scale ratio $z$ is equal to $R/r$. The crossing angle $\alpha$ is equal to the small increment $d\theta$ in the phase angle $\theta$. Traditional power spectrum analysis will resolve such a spiral flow trajectory as a continuum of eddies with progressive increase $d\theta$ in phase angle $\theta$.

with period length (Fig. 1.3). The peak periods $E_n$ in the dominant wave-bands is given by the relation

$$E_n = T_s(2 + \tau)\tau^n \qquad (1.4)$$

where $\tau$ is the *golden mean* equal to $(1 + \sqrt{5})/2$ (approximately equal to 1.618) and $T_s$, the primary perturbation time period, for example is 1 year, the annual cycle (summer to winter) of solar heating in a study of atmospheric inter-annual variability.

The peak periods $E_n$ are superimposed on a continuum background. For example, the most striking feature in climate variability on all time scales is the presence of sharp peaks superimposed on a continuous background (Ghil, 1994).

### 1.3.3.4  *Berry's phase in quantum systems*

The ratio $r/R$ also represents the increment $d\theta$ in phase angle $\theta$ (Eq. (1.3) and Fig. 1.4) and therefore the phase angle $\theta$ represents the variance (Selvam, 1990, 2005, 2007; Selvam and Fadnavis, 1998). Hence, when the logarithmic spiral is resolved as an eddy continuum in conventional spectral analysis, the increment in wavelength is concomitant with increase in phase. The angular turning, in turn, is directly proportional to the variance (Eq. (1.3)). Such a result that increments in wavelength and phase angle are related is observed in quantum systems and has been named *Berry's phase* (Berry, 1988). The relationship of angular turning of the spiral to intensity of fluctuations is seen in the tight coiling of the hurricane spiral cloud systems.

### 1.3.3.5  *Logarithmic spiral pattern underlying fractal fluctuation*

The overall logarithmic spiral flow structure is given by the relation

$$W = \frac{w_*}{k} \ln z \qquad (1.5)$$

where the constant $k$ is the steady state fractional volume dilution of large eddy by inherent turbulent eddy fluctuations. The constant $k$ is equal to

$$k = \frac{w_* r}{WR} = \frac{1}{\tau^2} \approx 0.382$$

$k$ is identified as the universal constant for deterministic chaos in fluid flows (Selvam, 1990, 2005, 2007; Selvam and Fadnavis, 1998). Since $k$ is less than half, the mixing with environmental air does not erase the signature of the dominant large eddy, but helps to retain its identity as a stable self-sustaining *soliton-like* structure. The mixing of environmental air assists in the upward and outward growth of the large eddy. The steady state emergence of fractal structures is therefore equal to

$$\frac{1}{k} \approx 2.62 \qquad (1.6)$$

Logarithmic wind profile relationship such as Eq. (1.5) is a long-established (observational) feature of atmospheric flows in the boundary layer, the constant $k$, called the *Von Karman*'s constant has the value equal to 0.38 as determined from observations (Wallace and Hobbs, 1977).

In Eq. (1.3), $W_{n+1}$ represents the standard deviation of eddy fluctuations at the $n+1^{\text{th}}$ step, since $W_{n+1}$ is computed as the instantaneous r.m.s. eddy perturbation amplitude with reference to the earlier step $W_n$ of eddy mean perturbation. For two successive stages of eddy growth starting from primary perturbation $w_*$ the ratio of the standard deviations $W_{n+1}$ and $W_n$ equal to $(n+1)/n$ is the standard deviation divided by the corresponding mean at the $n+1^{\text{th}}$ step of eddy growth and therefore represents the statistical normalized standard deviation $t$.

Denoting by $\sigma$ the standard deviation of eddy fluctuations at the reference level ($n = 1$) the standard deviations of eddy fluctuations for successive stages of eddy growth are given as integer multiple of $\sigma$, i.e., $\sigma$, $2\sigma$, $3\sigma$, etc., and correspond respectively to

Statistical normalized standard deviation $t = 0, 1, 2, 3,\ldots$ $\qquad (1.7)$

## 1.4  Universal Feigenbaum's Constants and Probability Density Distribution Function for Fractal Fluctuations

Selvam (1993, 2007) has shown that Eq. (1.1) represents the universal algorithm for deterministic chaos in dynamical systems and is expressed in terms of the universal *Feigenbaum's* (Feigenbaum, 1980) *constants a* and *d* as follows.

It is shown that *Feigenbaum's* constant *a* is equal to (Selvam, 1993, 2007)

$$a = \frac{W_2 R_2}{W_1 R_1} \tag{1.8}$$

In Eq. (1.8) the subscripts 1 and 2 refer to two successive stages of eddy growth. *Feigenbaum's* (Feigenbaum, 1980) constant *a* as defined above represents the steady state emergence of fractional *Euclidean* structures. Considering dynamical eddy growth processes, *Feigenbaum's* constant *a* also represents the steady state fractional outward mass dispersion rate and $a^2$ represents the energy flux into the environment generated by the persistent primary perturbation $W_1$. Considering both clockwise and counter-clockwise rotations, the total energy flux into the environment is equal to $2a^2$. In statistical terminology, $2a^2$ represents the variance of fractal structures for both clockwise and counter-clockwise rotation directions.

The steady state emergence of fractal structures in fluid flows is equal to $1/k$ ( $= \tau^2$) (Eq. (1.6)) and therefore the *Feigenbaum's* *constant a* is equal to

$$a = \tau^2 = \frac{1}{k} \approx 2.62 \tag{1.9}$$

The probability of occurrence $P_{tot}$ of fractal domain $W_1 R_1$ in the total larger eddy domain $W_n R_n$ in any (irrespective of positive or negative) direction is equal to

$$P_{tot} = \frac{W_1 R_1}{W_n R_n} = \tau^{-2n}$$

Therefore, the probability $P$ of occurrence of fractal domain $W_1R_1$ in the total larger eddy domain $W_nR_n$ in any one direction (either positive or negative) is equal to

$$P = \left(\frac{W_1R_1}{W_n R_n}\right)^{2n} = \tau^{-4n} \tag{1.10}$$

The *Feigenbaum*'s constant $d$ is shown to be equal to

$$d = \frac{W_2^4 R_2^3}{W_1^4 R_1^3} \tag{1.11}$$

Equation (1.11) represents the fractional volume intermittency of occurrence of fractal structures for each length step growth. *Feigenbaum*'s constant $d$ also represents the relative spin angular momentum of the growing large eddy structures as explained earlier.

Equation (1.1) may now be written as

$$2\frac{W^2 R^2}{w_*^2 (dR)^2} = \pi\frac{W^4 R^3}{w_*^4 (dR)^3} \tag{1.12}$$

In Eq. (1.12), $dR$ equal to $r$ represents the incremental growth in radius for each length step growth, i.e., $r$ relates to the earlier stage of eddy growth.

The *Feigenbaum*'s constant $d$ represented by $R/r$ is equal to

$$d = \frac{W^4 R^3}{w_*^4 r^3} \tag{1.13}$$

For two successive stages of eddy growth

$$d = \frac{W_2^4 R_2^3}{W_1^4 R_1^3} \tag{1.14}$$

From Eq. (1.1)

$$W_1^2 = \frac{2}{\pi} \frac{r}{R_1} w_*^2$$

$$W_2^2 = \frac{2}{\pi} \frac{r}{R_2} w_*^2$$

(1.15)

Therefore,

$$\frac{W_2^2}{W_1^2} = \frac{R_1}{R_2}$$

(1.16)

Substituting in Eq. (1.14)

$$d = \frac{W_2^4 R_2^3}{W_1^4 R_1^3} = \frac{W_2^2 W_2^2 R_2^3}{W_1^2 W_1^2 R_1^3} = \frac{R_1}{R_2} \frac{W_2^2 R_2^3}{W_1^2 R_1^3} = \frac{W_2^2 R_2^2}{W_1^2 R_1^2}$$

(1.17)

The *Feigenbaum*'s constant $d$ represents the scale ratio $R_2/R_1$ and the inverse of the *Feigenbaum*'s constant $d$ equal to $R_1/R_2$ represents the probability $(Prob)_1$ of occurrence of length scale $R_1$ in the total fluctuation length domain $R_2$ for the first eddy growth step as given in the following:

$$(Prob)_1 = \frac{R_1}{R_2} = \frac{1}{d} = \frac{W_1^2 R_1^2}{W_2^2 R_2^2} = \tau^{-4}$$

(1.18)

In general for the $n$th eddy growth step, the probability $(Prob)n$ of occurrence of length scale $R_1$ in the total fluctuation length domain $R_n$ is given as

$$(Prob)_n = \frac{R_1}{R_n} = \frac{W_1^2 R_1^2}{W_n^2 R_n^2} = \tau^{-4n}$$

(1.19)

The aforementioned equation for probability $(Prob)_n$ also represents, for the $n$th eddy growth step, the following statistical and dynamical quantities of the growing large eddy with respect to the initial perturbation domain: (i) the statistical relative variance of fractal structures, (ii) probability of occurrence of fractal domain in either positive or negative direction, and (iii) the inverse of $(Prob)_n$ represents the normalized fractal (fine scale) energy flux in the overall large scale eddy domain. Large-scale energy flux therefore occurs not in bulk, but in organized internal fine scale circulation structures identified as fractals.

Substituting the *Feigenbaum's constants a* and *d* defined earlier (Eqs. (1.8) and (1.11)), Eq. (1.12) can be written as

$$2a^2 = \pi d \qquad (1.20)$$

In Eq. (1.20) $\pi d$, the relative volume intermittency of occurrence contributes to the total variance $2a^2$ of fractal structures.

In terms of eddy dynamics, the aforementioned equation states that during each length step growth, the energy flux into the environment equal to $2a^2$ contributes to generate relative spin angular momentum equal to $\pi d$ of the growing fractal structures. Each length step growth is therefore associated with a factor of $2a^2$ equal to $2\tau^4$ ($\cong 13.7082$) increase in energy flux in the associated fractal domain. Ten such length step growths results in the formation of robust (self-sustaining) dominant bidirectional large eddy circulation $OR_0R_1R_2R_3R_4R_5$ (Fig. 1.3) associated with a factor of $20a^2$ equal to 137.08203 increase in eddy energy flux. This non-dimensional constant factor characterizing successive dominant eddy energy increments is analogous to the *fine structure* constant $\propto^{-1}$ (Ford, 1968) observed in atomic spectra, where the spacing (energy) intervals between adjacent spectral lines is proportional to the non-dimensional *fine structure* constant equal to approximately 1/137. Further, the probability of $n$th length step eddy growth is given by $a^{-2n}$ ($\cong 6.854^{-n}$) while the associated increase in eddy energy flux into the environment is equal to $a^{2n}$ ($\cong 6.854^n$). Extreme events occur for large number of length step growths $n$ with small probability of occurrence and are associated with large energy release in the fractal domain. Each length step growth is associated with one-tenth of *fine structure constant* energy increment equal to $2a^2$ ($\propto^{-1/10}$ $\cong 13.7082$) for bidirectional eddy circulation, or equal to one-twentieth of *fine structure constant* energy increment equal to $a^2$ ($\propto^{-1/20} \cong 6.854$) in any one direction, i.e., positive or negative. The energy increase between two successive eddy length step growths may be expressed as a function of $(a^2)^2$, i.e., proportional to the square of the *fine structure constant* $\propto^{-1}$.

In the spectra of many atoms, what appears with coarse observations to be a single spectral line proves, with finer observation, to be a group of two or more closely spaced lines. The spacing of these fine-structure lines relative to the coarse spacing in the spectrum is proportional to the square of *fine structure constant*, for which reason this combination is called the *fine-structure constant*. We now know that the significance of the *fine-structure constant* goes beyond atomic spectra (Ford, 1968). The atomic spectra therefore exhibit fractal-like structure within structure for the spectral lines.

### 1.4.1 *Same inverse power-law for probability distribution and power spectra of fractal fluctuations*

The power, i.e., variance spectra of fluctuations in fluid flows can now be quantified in terms of universal *Feigenbaum's constant a* as follows.

The steady state emergence of fractal structures is equal to the *Feigenbaum's constant a* (Eq. (1.8)). The relative variance of fractal structure which also represents the probability $P$ of occurrence of bidirectional fractal domain for each length step growth is then equal to $1/a^2$. The normalized variance $1/a^{2n}$ will now represent the statistical probability density for the $n$th step growth according to model predicted quantum-like mechanics for fluid flows. Model predicted probability density values $P$ are computed as

$$P = \frac{1}{a^{2n}} = \tau^{-4n} \tag{1.21}$$

or

$$P = \tau^{-4t} \tag{1.22}$$

In Eq. (1.22), $t$ is the normalized standard deviation (Eq. (1.7)).

The normalized variance and therefore the statistical probability distribution is represented by (from Eq. (1.10))

$$P = a^{-2t} \tag{1.23}$$

In Eq. (1.23), $P$ is the probability density corresponding to normalized standard deviation $t$. The probability density distribution of fractal fluctuations (Eq. (1.19)) is therefore the same as variance spectrum (Eq. (1.22)) of fractal fluctuations.

The graph of $P$ versus $t$ will represent the power spectrum. The slope $S$ of the power spectrum is equal to

$$S = \frac{dP}{dt} \approx -P \qquad (1.24)$$

The power spectrum therefore follows inverse power-law form, the slope decreasing with increase in $t$. Increase in $t$ corresponds to large eddies (low frequencies) and is consistent with observed decrease in slope at low frequencies in dynamical systems.

## 1.4.2 *Inverse power-law for fractal fluctuations close to Gaussian distribution*

The statistical distribution characteristics of fluctuations in natural phenomena follow normal distribution associated conventionally with random chance. The normal distribution is characterized by (1) the moment coefficient of skewness equal to zero, signifying symmetry and (2) the moment coefficient of kurtosis equal to 3 representing intermittency of turbulence on relative time scale.

In the following it is shown that the universal period doubling route to chaos growth phenomena in nature generates distribution characteristics very close to the observed statistical normal distribution parameters.

The successive length step growths generating the eddy continuum $OR_0R_1R_2R_3R_4R_5$ (Fig. 1.3) is analogous to the period doubling route to chaos (growth) and is initiated and sustained by the turbulent (fine scale) eddy acceleration $w_*$ which then propagates by the inherent property of the inertia of the medium. In the context of atmospheric turbulence, the statistical parameters, mean, variance, skewness and kurtosis represent respectively the net vertical velocity, intensity of turbulence, vertical momentum flux and intermittency of turbulence and are given respectively by $w_*$, $w_*^2$, $w_*^3$, $w_*^4$. By analogy, the perturbation speed $w_*$ (motion) per second of the medium sustained by its inertia represents the mass; $w_*^2$, the acceleration (or force); $w_*^3$, the momentum (or potential energy) and $w_*^4$, the spin angular momentum, since an eddy motion is inherently symmetric with bidirectional energy flow, the skewness factor $w_*^3$ is equal to zero for one complete eddy circulation thereby satisfying the law of conservation of momentum. The moment coefficient of kurtosis which represents the intermittency of turbulence is shown in the following to be slightly less than that for normal distribution value 3 for the period doubling route to chaos eddy growth phenomena. Selvam (2013) has shown that the probability distribution of fractal fluctuations is very close to the statistical normal distribution for values of normalized standard deviation $t \leq 2$ and gives appreciably larger probability values for larger standard deviations.

The numerical value of the fourth moment about the mean which represents the statistical intermittency of turbulence as well as the dynamical spin angular momentum of the eddy for the eddy growth process at Eq. (1.1) is computed in the following. From Eq. (1.5),

$$dW = \frac{w_*}{k} d \ln z$$

The fourth moment about the mean is given as

$$\left(\frac{dW}{w_*}\right)^4 = \left(\frac{d \ln z}{k}\right)^4 = \left(\frac{dz}{zk}\right)^4 = \left(\frac{r}{Rk}\right)^4 \tag{1.25}$$

The fourth moment about the mean $(\frac{dW}{W_*})^4$ represents the relative statistical coefficient of kurtosis and also the relative spin angular momentum for the eddy dynamical growth process. Organized eddy growth occurs for scale ratio equal to 10 and identifies the large eddy on whose envelope period doubling growth process occurs. Therefore, for a dominant eddy

$$\frac{r}{R} = \frac{1}{10}$$

From Eq. (1.1)
$$\frac{w_*}{W} = \sqrt{\frac{\pi}{2}\frac{R}{r}}$$

$k$, the steady state fractional volume dilution of large eddy turbulent eddy fluctuations is given as

$$k = \frac{w_* r}{W R} = \sqrt{\frac{\pi}{2} \frac{R}{r} \frac{r}{R}} = \sqrt{\frac{\pi}{2} \frac{r}{R}} = \sqrt{\frac{\pi}{20}} \approx 0.4$$

The computed value of $k$, the steady state fractional volume dilution of large eddy by turbulent eddy mixing and which is also the same as *von Karmen*'s constant in Eq. (1.5) is in agreement with experimentally measured value equal to about 0.4.

$(dz/z) = 1/2$ for one length growth by period doubling process since $z = dz + dz$ where $z$ is the length scale ratio equal to $R/r$. Therefore moment coefficient of kurtosis (Eq. (1.25)) is equal to

$$\left(\frac{dW}{w_*}\right)^4 = \left(\frac{r}{R k}\right)^4 = \frac{1}{2^4}\left(\frac{20 \times 7}{22}\right)^2 = 2.53 \qquad (1.26)$$

Moment coefficient of kurtosis (Eq. (1.26)) equal to 2.53 is less than that for the normal distribution value 3.0. Selvam (2013) has shown that probability distribution of fractal fluctuations is close to normal distribution for small to moderate fluctuations (up to 2 standard deviations) but exhibits a fat long tail for larger fluctuations. Period doubling growth phenomena results in a 2.53 times increase in the spin angular momentum of the large eddy for each period doubling sequence. Period doubling at constant pump frequency involves eddy length step growth $dR$ on either side of the primary turbulent eddy length $dR$.

### 1.4.3 *Fat long tail for probability distribution of fractal fluctuations*

Selvam (2012a, 2012b, 2013) has shown (i) Eq. (1.22) gives the theoretical probability density $P$ for $t < -1$ and $t > 1$ (ii) the theoretical probability density $P$ for $t > -1$ and $t < 1$ is given in terms of $k$, the steady state fractional volume dilution of large eddy by turbulent eddy fluctuations (see Section 1.4) as

$$P = a^{-2k}$$

**Table 1.2:** Model predicted and statistical normal probability density distributions.

| Growth step | Normalized deviation | Cumulative probability densities (%) | |
|---|---|---|---|
| n | t | Model predicted $P = \tau^{-4t}$ | Statistical normal distribution |
| 1 | 1 | 14.5898 | 15.8655 |
| 2 | 2 | 2.1286 | 2.2750 |
| 3 | 3 | 0.3106 | 0.1350 |
| 4 | 4 | 0.0453 | 0.0032 |
| 5 | 5 | 0.0066 | $\approx 0.0$ |

The theoretical distribution is (i) close to statistical normal distribution for $-2 < t < 2$ values and (ii) the theoretical distribution gives higher probability values for $t < -2$ and $t > 2$ (see Table 1.2).

The model predicted $P$ values corresponding to normalized deviation $t$ values less than 2 are close to the corresponding statistical normal distribution values while the $P$ values are noticeably larger for normalized deviation $t$ values greater than 2 (Table 1.2 and Fig. 1.5) and may explain the reported *fat tail* for probability distributions of various physical parameters (Buchanan, 2004).

The model predicted $P$ values plotted on a linear scale ($Y$-axis) shows close agreement with the corresponding statistical normal probability values as seen in Fig. 1.5 (left side). The model predicted $P$ values plotted on a logarithmic scale ($Y$-axis) shows *fat tail* distribution for normalized deviation $t$ values greater than 2 as seen in Fig. 1.5 (right side).

### 1.4.4 *Power spectra of fractal fluctuations*

It was shown at Section 1.4 that the same inverse power-law represents probability density distribution (Eq. (1.19)) and power spectra (Eq. (1.23)) of fractal fluctuations. Therefore, the conventional power spectrum plotted as the variance versus the frequency in log–log scale will now represent the eddy probability density on logarithmic scale versus the standard

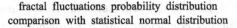

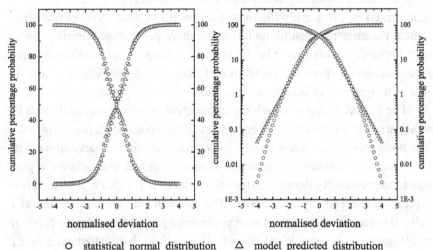

○ statistical normal distribution  △ model predicted distribution

**Figure 1.5:** Comparison of statistical normal distribution and computed (theoretical) probability density distribution. The same figure is plotted on the right side with logarithmic scale for the probability axis (*Y*-axis) to show clearly that for normalized deviation *t* values greater than 2 the computed probability densities are greater than the corresponding statistical normal distribution values.

deviation of the eddy fluctuations on linear scale since the logarithm of the eddy wavelength represents the standard deviation, i.e., the r.m.s. value of eddy fluctuations (Eq. (1.5)). The r.m.s. value of eddy fluctuations can be represented in terms of probability density distribution as follows. A normalized standard deviation *t* = 0 corresponds to cumulative percentage probability density equal to 50 for the mean value of the distribution. Since the logarithm of the wavelength represents the r.m.s. value of eddy fluctuations the normalized standard deviation *t* is defined for the eddy energy as

$$t = \frac{\log L}{\log T_{50}} - 1 \qquad (1.27)$$

In Eq. (1.27), *L* is the time period (or wavelength) and $T_{50}$ is the period up to which the cumulative percentage contribution to total variance is equal to 50 and $t = 0$. Log$T_{50}$ also represents the mean value for the r.m.s. eddy fluctuations and is consistent with the concept of the mean level represented by r.m.s. eddy fluctuations. Spectra of time series of meteorological parameters when plotted as cumulative percentage contribution to total variance versus $t$ have been shown to follow closely the model predicted universal spectrum (Selvam and Fadnavis, 1998) which is identified as a signature of quantum-like chaos.

The period $T_{50}$ up to which the cumulative percentage contribution to total variance is equal to 50 is computed from model concepts as follows. The power spectrum, when plotted as normalized standard deviation $t$ (Eq. (1.7)) versus cumulative percentage contribution to total variance represents the probability density distribution (Section 1.4.1), i.e., the variance represents the probability density. The normalized standard deviation $t$ value 0 corresponds to cumulative percentage probability density $P$ equal to 50, same as for statistical normal distribution characteristics. Since $t$ represents the eddy growth step $n$ (Eq. (1.7)), the dominant period $T_{50}$ up to which the cumulative percentage contribution to total variance is equal to 50 is obtained from Eq. (1.4) for value of $n$ equal to 0. In the present study of periodicities in frequency distribution of genomic DNA base at unit spacing intervals, the primary perturbation time period $T_s$ is equal to unit number class interval and $T_{50}$ is obtained in terms of unit number class interval as

$$T_{50} = T_s(2 + \tau)\tau^0 \approx 3.6 \tag{1.28}$$

Genomic DNA base with spacing intervals up to 3.6 or approximately 4 contribute up to 50% to the total variance. This model prediction is in agreement with computed values of $T_{50}$ for the different DNA sequences analyzed and discussed in Chapters 3–8.

The general systems theory concepts are equivalent to Boltzmann's postulates and the *Boltzmann distribution* (for molecular energies) with the inverse power-law expressed as a function of the *golden mean* is the universal probability distribution function for (the amplitude as well as variance) of observed fractal fluctuations which corresponds closely to statistical normal distribution for moderate amplitude fluctuations and exhibit a fat long tail for hazardous extreme events in dynamical systems

(Selvam, 2014a, 2014b). The general systems theory model predictions given above are based on classical statistical physical concepts and satisfy the principle of maximum entropy production for dynamical system in steady-state equilibrium (Selvam, 2013).

## Acknowledgement

The author is grateful to Dr. A. S. R. Murty for encouragement.

## References

Aharony, A. and Feder, J., eds. (1989). *Fractals in Physics*: *Essays in Honour of Benoit B. Mandebrot* (North-Holland, Amsterdam).

Andriani, P. and McKelvey, B. (2007). Beyond Gaussian averages: Redirecting management research toward extreme events and power-laws, *J. Int. Bus. Stud.*, 38, pp. 1212–1230.

Auerbach, F. (1913). Das Gesetz Der Bevolkerungskoncentration, *Petermanns Geogr. Mitt*, 59, pp. 74–76.

Bak, P. C., Tang, C., and Wiesenfeld, K. (1988). Self-organized criticality, *Phys. Rev.* A, 38, pp. 364–374.

Berry, M. V. (1988). The geometric phase, *Sci. Amer.*, Dec., pp. 26–32.

Bouchaud, J. P., Sornette, D., Walter, C., and Aguilar, J. P. (1998). Taming large events: Optimal portfolio theory for strongly fluctuating assets, *Int. J. Theor. Appl. Finance*, 1(1), pp. 25–41.

Buchanan, M. (2004). Power-laws and the new science of complexity management, *Strategy and Business Issue*, 34, pp. 70–79.

Clauset, A., Shalizi, C. R., and Newman, M. E. J. (2007). Power-law distributions in empirical data, *SIAM Rev.*, 51(4), pp. 661–703.

Cronbach, L. (1970). *Essentials of Psychological Testing* (Harper & Row, New York).

Curl, R. F. and Smalley, R. E. (1991). Fullerenes, *Scientific American* (Indian Edition) 3, pp. 32–41.

Davenas, E., Beauvais, F., Amara, J., Oberbaum, M., Fortner, O., Belon, P., Sainte-Laudy, J., Robinzon, B., Miadonna, A., Tedeschi, A., Pomeranz, B., and Benvieste, J. (1988). Human basophil degranulation triggered by very dilute antiserum against IgE, *Nature*, 333, pp. 816–818.

Estoup, J. B. (1916). *Gammes Stenographiques* (Institut Stenographique de France, Paris).

Fama, E. F. (1965). The behavior of stock-market prices, *J. Bus.*, 38, pp. 34–105.

Feigenbaum, M. J. (1980). Universal behavior in nonlinear systems, *Los Alamos Sci.*, 1980, 1, pp. 4–27.

Ford, K. W. (1968) *Basic Physics* (Blaisdell Publishing, Waltham).

Ghil, M. (1994). Cryothermodynamics: The chaotic dynamics of paleoclimate, *Physica D*, 77, pp. 130–159.

Goertzel, T. and Fashing, J. (1981). The myth of the normal curve: A theoretical critique and examination of its role in teaching and research, *Humanity and Society* 5, pp. 14–31; reprinted in *Readings in Humanist Sociology*, General Hall, 1986. http://crab.rutgers.edu/~goertzel/normalcurve.htm 4/29/2007.

Greene, W. H. (2002) *Econometric Analysis*, 5th Ed. (Prentice-Hall, Englewood Cliffs, New Jersey).

Grossing, G. (1989). Quantum systems as order out of chaos phenomena, *Il Nuovo Cimento*, 103B, pp. 497–510.

Gutenberg, B. and Richter, R. F. (1944). Frequency of earthquakes in California, *Bull. Seis. Soc. Amer.*, 34, pp. 185–188.

Janssen, T. (1988). Aperiodic crystals: A contradictio in terminals? *Phys. Rep.*, 168(2), pp. 55–113.

Jean, R. V. (1992a). Nomothetical modelling of spiral symmetry in biology. In *Fivefold Symmetry*, ed. Hargittai, I. (World Scientific, Singapore).

Jean, R. V. (1992b). On the origins of spiral symmetry in plants. In *Spiral Symmetry*, eds. Hargittai, I. and Pickover, C. A. (World Scientific, Singapore).

Jean, R. V. (1994). *Phyllotaxis: A Systemic Study in Plant Morphogenesis* (Cambridge University Press, New York).

Liebovitch, L. S., and Scheurle, D. (2000). Two lessons from fractals and chaos, *Complexity*, 5(4), pp. 34–43.

Maddox, J. (1988). Licence to slang Copenhagen? *Nature*, 332, p. 581.

Maddox, J. (1993). Can quantum theory be understood? *Nature*, 361, p. 493.

Mandelbrot, B. B. (1963). The variation of certain speculative prices, *J. Bus.*, 36(4), pp. 394–419.

Mandelbrot, B. B. (1975). On the geometry of homogenous turbulence with stress on the fractal dimension of the iso-surfaces of scalars, *J. Fluid Mech.*, 72, pp. 401–416.

Mandelbrot, B. B. (1977). *Fractals: Form, Chance and Dimension* (Freeman, San Francisco).

Mandelbrot, B. B. (1983). *The Fractal Geometry of Nature* (W. H. Freeman and Company).

Mandelbrot, B. B. and Hudson, R. (2004). *The (Mis)behavior of Markets: A Fractal View of Risk, Ruin, and Reward* (Basic Books, Cambridge).

Milotti, E. (2002). $1/f$ noise: A pedagogical review. http://arxiv.org/abs/physics/0204033.

Montroll, E. and Shlesinger, M. (1984). On the wonderful world of random walks. In *Nonequilibrium Phenomena II, from Stochastic to Hydrodynamics*, eds. Lebowitz, J. L. and Montroll, E. W. (North Holland, Amsterdam), pp. 1–121.

Omori, F. (1894). On the aftershocks of earthquakes, *J. Coll. Sci.*, 7, pp. 111–200.

Pareto, V. (1897) *Cours d'Economie Politique* (Rouge, Lausanne and Paris).

Pearson, K. (1900). On the criterion that a given system of deviations from the probable in the case of a correlated system of variables is such that it can be reasonably supposed to have arisen from random sampling, *Philosophical Magazine*, 5th Series, Vol. L, pp. 157–175.

Pearson, H. (2004). "Junk" DNA reveals vital role, *Nature Science update* 7 May. https://doi.org/10.1038/news040503-9.

Phillips, T. (2005). *The Mathematical Uncertainty Principle*, Monthly Essays on Mathematical Topics, November 2005, American Mathematical Society. http://www.ams.org/featurecolumn/archive/uncertainty.html.

Planat, M. (2003). *The Nature of Time: Geometry, Physics and Perception*, eds. Buccheri, R., Saniga, M., and Stuckey, W. M. (Springer, Berlin).

Planat, M., Rosu, H., and Perrine, S. (2002). Ramanujan sums for signal processing of low-frequency noise, *Phys. Rev. E*, 66(5), pp. 056128(1–7).

Poincare, H. (1892). *Les Methodes Nouvelle de la Mecannique Celeste* (Gautheir-Villars, Paris).

Rae, A. (1988). *Quantum-Physics: Illusion or Reality?* (Cambridge University Press, New York).

Richardson, L. F. (1960). The problem of contiguity: An appendix to statistics of deadly quarrels. In *General Systems — Year Book of the Society for General Systems Research*, eds. Von Bertalanffy, L., and Rapoport, A., pp. 139–187 (Ann Arbor, Michigan).

Riley, K. F., Hobson, M. P., and Bence, S. J. (2006). *Mathematical Methods for Physics and Engineering*, 3rd Ed. (Cambridge University Press, Cambridge).

Riste, T. and Sherrington, D., eds. (1991). *Spontaneous Formation of Space-Time Structures and Criticality* (Nato Science Series C), (Springer-Verlag, London).

Ruhla, C. (1992). *The Physics of Chance* (Oxford University Press, Oxford), p. 217.

Schroeder, M. (1990). *Fractals, Chaos and Power-Laws* (W. H. Freeman and Co., New York).

Selvam, A. M. (1990). Deterministic chaos, fractals and quantum-like mechanics in atmospheric flows, *Can. J. Phys.*, 68, pp. 831–841. http://xxx.lanl.gov/html/physics/0010046.

Selvam, A. M. (1993). Universal quantification for deterministic chaos in dynamical systems, *Appl. Math. Model.*, 17, pp. 642–649. http://xxx.lanl.gov/html/physics/0008010.

Selvam, A. M. and Fadnavis, S. (1998). Signatures of a universal spectrum for atmospheric inter-annual variability in some disparate climatic regimes, *Meteor. Atmos. Phys.*, 66, pp. 87–112. http://xxx.lanl.gov/abs/chao-dyn/9805028.

Selvam, A. M. and Suvarna Fadnavis, S. (1999a). Superstrings, cantorian-fractal space-time and quantum-like chaos in atmospheric flows, *Chaos, Solit. Fractals*, 10(8), pp. 1321–1334. http://xxx.lanl.gov/abs/chao-dyn/9806002.

Selvam, A. M. and Fadnavis S. (1999b). Cantorian fractal space-time, quantum-like chaos and scale relativity in atmospheric flows, *Chaos, Solit. Fractals*, 10(9), pp. 1577–1582. http://xxx.lanl.gov/abs/chao-dyn/9808015.

Selvam, A. M. (2001a). Quantum-like chaos in prime number distribution and in turbulent fluid flows, *APEIRON*, 8(3), pp. 29–64.

Selvam, A. M. (2001b). Signatures of quantum-like chaos in spacing intervals of non-trivial *Riemann* Zeta zeros and in turbulent fluid flows, *APEIRON*, 8(4), pp. 10–40. http://xxx.lanl.gov/html/physics/0102028.

Selvam, A. M. (2002). Quantum-like chaos in the frequency distributions of the bases A, C, G, T in *Drosophila* DNA, *APEIRON*, 9(4), pp. 103–148. http://redshift.vif.com/JournalFiles/V09NO4PDF/V09N4sel.pdf.

Selvam, A. M. (2003). Signatures of quantum-like chaos in Dow Jones Index and turbulent fluid flows, *APEIRON*, 10, pp. 1–28. http://arxiv.org/html/physics/0201006, http://redshift.vif.com/JournalFiles/V10NO4PDF/V10N4SEL.PDF.

Selvam, A. M. (2004). Quantum-like chaos in the frequency distributions of bases A, C, G, T in Human chromosome1 DNA, *APEIRON*, 11(3), pp. 134–146. http://redshift.vif.com/JournalFiles/V11NO3PDF/V11N3SEL.PDF.

Selvam, A. M. (2005). A general systems theory for chaos, quantum mechanics and gravity for dynamical systems of all space-time scales, *Electromagnetic Phenomena*, 52(15), pp. 160–176. arXiv:physics/0503028 (physics.gen-ph).

Selvam, A. M. (2007). *Chaotic Climate Dynamics* (Luniver Press, United Kingdom).

Selvam, A. M. (2009). Fractal fluctuations and statistical normal distribution, *Fractals*, 17(3), pp. 333–349.

Selvam, A. M. (2011). Signatures of universal characteristics of fractal fluctuations in global mean monthly temperature anomalies, *J. Sys. Sci. Complexity*, 24, pp. 14–38.

Selvam, A. M. (2012a). Universal spectrum for atmospheric suspended particulates: Comparison with observations, *Chaos Complexity Lett.*, 6(3), pp. 1–43. http://arxiv.org/abs/1005.1336.

Selvam, A. M. (2012b). Universal spectrum for atmospheric aerosol size distribution: Comparison with pcasp-b observations of vocals 2008, *Nonlinear Dyn. Syst. Theory*, 12(4), pp. 397–434. http://arxiv.org/abs/1105.0172.

Selvam, A. M. (2013). Scale-free universal spectrum for atmospheric aerosol size distribution for Davos, Mauna Loa and Izana, *Int. J. Bifurc. Chaos*, 23(2), pp. 1350028 (13 pages).

Selvam, A. M. (2014a). Universal inverse power-law distribution for temperature and rainfall in the UK region, *Dyn. Atmos. Oceans*, 66, pp. 138–150.

Selvam, A. M. (2014b). Universal characteristics of fractal fluctuations in prime number distribution, *Int. J. Gen. Syst.*, 43(8), pp. 828–863.

Sornette, D., Johansen, A., and Bouchaud, J.-P. (1995). Stock Market Crashes, Precursors and Replicas. http://xxx.lanl.gov/pdf/cond-mat/9510036.

Sornette, D. (2007). *Probability Distributions in Complex Systems*. http://arxiv.org/abs/0707.2194v1.

Steinbock, O., Zykov, V., and Muller, S. C. (1993). Control of spiral-wave dynamics in active media by periodic modulation of excitability, *Nature*, 366, pp. 322–324.

Steinhardt, P. (1997). Crazy crystals, *New Scientist*, 25 January, pp. 32–35.

Stevens, P. S. (1974). *Patterns in Nature* (Little, Brown and Co. Inc., Boston).

Tarasov, L. (1986). *This Amazingly Symmetrical World* (Mir Publishers, Moscow).

Townsend, A. A. (1956). *The Structure of Turbulent Shear Flow*, 2nd Ed., (Cambridge University Press, London), pp. 115–130.

Wallace, J. M. and Hobbs, P. V. (1977). *Atmospheric Science: An Introductory Survey* (Academic Press, New York).

Zipf, G. K. (1949). Human Behavior and the Principle of Least Effort (Hafner, New York).

# Chapter 2

# Nonlinear Dynamics, Chaos and Self-organized Criticality

## 2.1 Introduction

Irregular (nonlinear) fluctuations on all scales of space and time are generic to dynamical systems in nature such as fluid flows, atmospheric weather patterns, heart beat patterns, stock market fluctuations, the DNA sequence of letters A, C, G and T, etc. Mandelbrot (1977) coined the name *fractal* for the non-Euclidean geometry of such fluctuations which have fractional dimension, for example, the rise and subsequent fall with time of the Dow Jones Index or rainfall that traces a zig-zag line in a two-dimensional plane and therefore has a *fractal* dimension greater than one but less than two. Mathematical models of dynamical systems are nonlinear and finite precision computer realizations exhibit sensitive dependence on initial conditions resulting in chaotic solutions, identified as deterministic chaos. *Nonlinear dynamics and chaos* is now (since 1980s) an area of intensive research in all branches of science (Gleick, 1987). The *fractal* fluctuations exhibit scale invariance or self-similarity manifested as the widely documented (Bak *et al.*, 1988; Bak and Chen, 1989, 1991; Schroeder, 1991; Stanley, 1995; Buchanan, 1997) inverse power-law form for power spectra of space-time fluctuations identified as *self-organized criticality* by Bak *et al.* (1987). The power-law is a distinctive experimental signature seen in a wide variety of complex systems. In economy it goes by the name fat tails, in physics it is referred to as critical

37

fluctuations, in computer science and biology it is the edge of chaos, and in demographics it is called *Zipf's* law (Newman, 2000). Power-law scaling is not new to economics. The power-law distribution of wealth discovered by *Vilfredo Pareto* (1848–1923) in the 19th century (Eatwell *et al.*, 1991) predates any power-laws in physics (Farmer, 1999). One of the oldest scaling laws in geophysics is the *Omori law* (Omori, 1895). It describes the temporal distribution of the number of aftershocks, which occur after a larger earthquake (i.e., main-shock) by a scaling relationship. The other basic empirical seismological law, the *Gutenberg–Richter law* (Gutenberg and Richter, 1944) is also a scaling relationship, and relates intensity to its probability of occurrence (Hooge *et al.*, 1994). Time series analyses of global market economy also exhibits power-law behaviour (Bak *et al.*, 1992; Mantegna and Stanley, 1995; Sornette *et al.*, 1995; Chen, 1996a, 1996b; Stanley *et al.*, 1996c; Feigenbaum and Freund, 1997a, 1997b; Gopikrishnan *et al.*, 1999; Plerou *et al.*, 1999; Stanley *et al.*, 2000; Feigenbaum, 2001a, 2001b) with possible *multifractal* structure (Farmer, 1999) and has suggested an analogy to fluid turbulence (Ghashghaie *et al.*, 1996; Arneodo *et al.*, 1998). Sornette *et al.* (1995) conclude that the observed power law represents structures similar to *Elliott waves* of technical analysis first introduced in the 1930s. It describes the time series of a stock price as made of different waves; these waves are in relation to each other through the *Fibonacci* series. Sornette *et al.* (1995) speculate that *Elliott waves* could be a signature of an underlying critical structure of the stock market. Incidentally, the *Fibonacci* series represent a *fractal* tree-like branching network of self-similar structures (Stewart, 1992). The commonly found shapes in nature are the *helix* and the *dodecahedron* (Muller and Beugholt, 1996), which are signatures of self-similarity underlying *Fibonacci* numbers. The general systems theory presented in this chapter shows (Section 1.3) that *Fibonacci* series underlies *fractal* fluctuations on all space-time scales.

Historically, basic similarity in the branching (*fractal*) form underlying the individual leaf and the tree as a whole was identified more than three centuries ago in botany (Arber, 1950). The branching (bifurcating) structure of roots, shoots, veins on leaves of plants, etc., have similarity in form to branched lighting strokes, tributaries of rivers, physiological networks of blood vessels, nerves and ducts in lungs, heart, liver, kidney, brain, etc. (Freeman, 1987, 1990; Goldberger *et al.*, 1990; Jean, 1994). Such seemingly complex network structure is again associated with *Fibonacci* numbers seen in the exquisitely ordered beautiful patterns in flowers and arrangement of leaves in the plant kingdom (Jean, 1994; Stewart, 1995). The identification of physical mechanism for the spontaneous generation of mathematically precise, robust spatial pattern formation in plants will have direct applications in all other areas of science (Mary Selvam, 1998). The importance of scaling concepts were recognized nearly a century ago in biology and botany where the dependence of a property $y$ on size $x$ is usually expressed by the *allometric* equation $y = ax^b$ where $a$ and $b$ are constants (Thompson, 1963; Strathmann, 1990; Jean, 1994; Stanley *et al.*, 1996b). This type of scaling implies a hierarchy of substructures and was used by *D'Arcy Thompson* for scaling anatomical structures, for example, how proportions tend to vary as an animal grows in size (West, 1990a). Thompson (1963, first published in 1917) in his book *On Growth and Form* has dealt extensively with similitude principle for biological modelling. Rapid advances have been made in recent years in the fields of biology and medicine in the application of scaling (*fractal*) concepts for description and quantification of physiological systems and their functions (Goldberger *et al.*, 1990; West, 1990a, 1990b; Deering and West, 1992; Skinner, 1994; Stanley *et. al.*, 1996b). In meteorological theory, the concept of self-similar fluctuations was identified and introduced in the description of turbulent flows by Richardson (1965, originally published in 1922; see also Richardson, 1960), Kolmogorov (1941, 1962), Mandelbrot (1975), Kadanoff (1996) and others (see Monin and Yaglom (1975) for a review).

*Self-organized* criticality implies long-range space-time correlations or non-local connections in the spatially extended dynamical system. The physics underlying *self-organized criticality* is not yet identified. Prediction of the future evolution of the dynamical system requires precise quantification of the observed *self-organized criticality*. The author has developed a general systems theory (Capra, 1996), which predicts the observed *self-organized criticality* as a signature of quantum-like chaos in the macroscale dynamical system (Mary Selvam, 1990; Mary Selvam *et al.*, 1992; Selvam and Fadnavis, 1998). The model also provides universal and unique quantification for the observed *self-organized criticality* in terms of the golden mean ($\approx$1.618); the model predicted distribution being close to statistical normal distribution for values of normalized standard deviation (mean/standard deviation) $t \leq 2$ on either side of the mean.

Long-range space-time correlations, manifested as the self-similar fractal geometry to the spatial pattern, concomitant with inverse power law form for power spectra of space-time fluctuations are generic to spatially extended dynamical systems in nature and are identified as signatures of self-organized criticality. A representative example is the self-similar fractal geometry of *His–Purkinje* system whose electrical impulses govern the inter-beat interval of the heart. The spectrum of inter-beat intervals exhibits a broadband inverse power law form $f^{-\alpha}$ where $f$ is the frequency and $\alpha$ the exponent. Self-organized criticality implies non-local connections in space and time, i.e., long-term memory of short-term spatial fluctuations in the extended dynamical system that acts as a unified whole communicating network.

*Nonlinear dynamics and chaos*, a multidisciplinary area of intensive research in recent years (since 1980s) has helped identify universal characteristics of spatial patterns (forms) and temporal fluctuations (functions) of disparate dynamical systems in nature. Examples of dynamical systems, i.e., systems which change with time include biological (living) neural networks of the human brain which responds as a unified whole to a multitude of input signals and the non-biological (non-living) atmospheric flow structure which exhibits teleconnections, i.e., long-range space-time correlations. Spatially extended dynamical systems in nature exhibit self-similar fractal geometry to the spatial pattern. The sub-units of self-similar structures resemble the whole in shape. The name "fractal" coined by Mandelbrot (1977) indicates non-Euclidean or fractured (broken) Euclidean structures. Traditional Euclidean geometry discusses only three-, two- and one-dimensional objects, representative examples being sphere, rectangle and straight lines, respectively. Objects in nature have irregular non-Euclidean shapes, now identified as fractals and the fractal dimension $D$ is given as $D = d\log M/d\log R$, where $M$ is the mass contained within a distance $R$ from a point within the extended object. A constant value for $D$ indicates uniform stretching on logarithmic scale. Objects in nature, in general exhibit multifractal structure, i.e., the fractal dimension $D$ varies with length scale $R$. The fractal structure of physiological systems has been identified (Goldberger *et al.*, 1990, 2002; West, 1990, 2004). The global atmospheric cloud cover pattern also exhibits self-similar fractal geometry (Lovejoy and Schertzer, 1986). Self-similarity implies long-range spatial correlations, i.e., the larger scale is a magnified version of the smaller scale with enhancement of fine structure. Self-similar fractal structures in nature support functions, which fluctuate on all scales of time. For example, the neural network of the human brain responds to a multitude of sensory inputs on all scales of time with long-term memory update and retrieval for appropriate global response to local input signals. Fractal architecture to the spatial pattern enables integration of a multitude of signals of all space-time scales so that the dynamical system responds as a unified whole to local stimuli, i.e., short-term fluctuations are carried as internal structure to long-term fluctuations. Fractal networks therefore function as dynamic memory storage devices, which integrate short-term fluctuations into long-period fluctuations. The irregular (nonlinear) variations of fluctuations in dynamical systems are therefore broadband because of coexistence of fluctuations of all scales. Power spectral analysis (MacDonald, 1989) is conventionally used to resolve the periodicities (frequencies) and their amplitudes in time series data of fluctuations. The power spectrum is plotted on log–log scale as the intensity represented by variance (amplitude squared) versus the period (frequency) of the component periodicities. Dynamical systems in nature exhibit inverse power law form $1/f^{\alpha}$ where $f$ is the frequency (1/period) and $\alpha$ the exponent for the power spectra of space-time fluctuations indicating self-similar fluctuations on all space-time scales, i.e., long-range space-time correlations. The amplitudes of short term and long term are related by a scale factor alone, i.e., the space-time fluctuations exhibit scale invariance or long-range space-time correlations, which are independent of the exact details of dynamical mechanisms underlying the fluctuations at different scales. The universal characteristics of spatially extended dynamical systems, namely, the fractal structure to the space-time fluctuation pattern and inverse power law form for power spectra of space-time fluctuations are identified as signatures of self-organized criticality (Bak *et al.*, 1988). Self-organized criticality implies non-local connections in space and time in real world dynamical systems.

Surprisingly, such long-range space-time correlations had been earlier identified (Gleick, 1987) as sensitive dependence on initial conditions of finite precision computer realizations of nonlinear mathematical models of dynamical systems and named "deterministic chaos". Deterministic chaos is therefore a signature of self-organized criticality in computed model solutions.

It has not been possible to identify the exact mechanism underlying the observed universal long-range space-time correlations in natural dynamical systems and in computed solutions of model dynamical systems. The physical mechanisms responsible for self-organized criticality should be independent of the exact details (physical, chemical, physiological, biological, computational system, etc.) of the dynamical system so as to be universally applicable to all dynamical systems (real and model).

## 2.2 The DNA Molecule and Heredity

Heredity in living organisms is determined by a long complex chemical molecule called deoxyribonucleic acid (DNA). The units of heredity, the genes are parts of the DNA molecule situated along the length of the chromosomes inside the nucleus of the cell.

Historically, Watson and Crick (1953) put together all the experimental data concerning DNA and decided that the only structure that fitted all the facts was the double helix and postulated that DNA is composed of two ribbon-like "backbones" composed of alternating deoxyribose and phosphate molecules.

A simplified picture of the molecule of DNA may be visualized to consist of two long backbones with projections sticking out from them at right angles rather like a ladder with its two upright sides and its rungs. The backbones are made up of two simple chemicals arranged alternately — *sugar — phosphate — sugar — phosphate* — all along the way. The projections are the four units or "letters" of the code; they are four chemical bases called *guanine, cytosine, adenine* and *thymine* — G, C, A and T. These four bases are arranged in a specific sequence, which constitutes the genetic code. The DNA molecule actually consists not of a single thread, but of two helical threads wound around each other — a double helix. The two DNA chains run in opposite directions and are coiled around each other with the bases facing one another in pairs. Only specific pairs of bases can be linked together, T always pairs with A, and G with C (Claire, 1964; Bates and Maxwell, 1993). The amount of A is the same as the amount of T, while the amount of G is the same as the amount of C. These are now known as Chargaff ratios (Gribbin, 1985; Alcamo, 2001).

The AT "base pair" forms with two hydrogen bonds and GC forms a base pair with three hydrogen bonds. Consequently, one may characterize AT base pairs as weak bases and GC base pairs as strong bases. Therefore, DNA sequences can be transformed into sequences of weak W (A or T) and strong S (G or C) bases (Li and Holste, 2004). The SW mapping rule is particularly appropriate to analyze genome-wide correlations; this rule corresponds to the most fundamental partitioning of the four bases into their natural pairs in the double helix (G+C, A+T). The composition of base pairs, or GC level, is thus a strand-independent property of a DNA molecule and is related to important physico-chemical properties of the chain (Bernaola-Galvan *et al.*, 2002). The C+G content (isochore) studies have been done earlier (Bernardi *et al.*, 1985; Ikemura, 1985; Ikemura, and Aota, 1988; Bernardi, 1989). The full story of how DNA really functions is not merely what is written on the sequence of base pairs. The DNA functions involve information transmission over many length scales ranging from a few to several hundred nanometers (Ball, 2000).

What distinguishes one type of cell from another and one organism from another is the protein, which it contains. And it is DNA which dictates to the cell how many and what types of protein it shall make. Twenty different chemicals called amino acids in different sets of combinations form the proteins. The sequence of bases along each DNA molecule in the chromosome determines the sequence of amino acids along each of the proteins. It takes a sequence of *three* bases, the codon, to identify one amino acid. The order in which these bases recur within a particular gene in the helix corresponds to the information needed to build that gene's particular protein (Claire, 1964; Leone, 1992; Ball, 2000).

## 2.3 Junk DNA

The four chemical components called bases, namely, Adenine (A), thymine (T); cytosine (C) and guanine (G) which hold the double-stranded DNA molecule together form the "code of life"; there are close to three billion base pairs in mammals such as humans and rodents. Written in the DNA of these animals are 25,000–30,000 genes which cells use as templates to start the production of proteins; these sophisticated molecules build and maintain the body. According to the traditional viewpoint, the really crucial things were genes, which code for proteins — the "building blocks of life". A few other sections that regulate gene function were also considered useful. The rest was thought to be excess baggage — or "junk" DNA. But new findings suggest this interpretation may be wrong. Comparison of genome sequences of man, mouse and rat and also chicken, dog and fish sequences show that several great stretches of DNA were identical across the species which shared an ancestor about 400 million years ago. These "ultra-conserved" regions do not appear to code for protein, but obviously are of great importance for survival of the animal. Nearly a quarter of the sequences overlap with genes and may help in protein production. The conserved elements that do not actually overlap with genes tend to cluster next to genes that play a role in embryonic development (Kettlewell, 2004; Check, 2006).

A number of the elements in the noncoding portion of the genome may be central to the evolution and development of multicellular organisms. Understanding the functions of these non-protein-coding sequences, not just the functions of the proteins themselves, will be vital to understanding the genetics, biology and evolution of complex organisms (Taft and Mattick, 2003).

The genes of higher organisms are seldom "recorded" in the chromosomes intact, but are scattered in fragmentary fashion along a stretch of DNA, broken up by chunks of DNA which seem at first sight to carry no message at all. All the useless or "junk" DNA, the intervening sequences are known as *introns*. The pieces of DNA carrying genetic code are called *exons*. The codons, 64 in number are distributed over the coding parts of the DNA sequences. It is well known that the coding regions are translated into proteins. The non-coding parts are presumed important in regulatory and promotional activities. The biologically meaningful structures in noncoding regions are not known (Gribbin, 1985; Guharay *et al.*, 2000; Clark, 2001; Som *et al.*, 2001). Understanding genetic defects will make it easier to treat them (Watson, 1997).

During the late 1960s papers began appearing that showed eukaryotic DNA contained large quantities of repetitive DNA which did not appear to code for proteins. By the early 1970s, the term "junk DNA" had been coined to refer to this non-coding DNA. Junk DNA seemed like an appropriate term for DNA cluttering up the genome while contributing in no way to the protein coding function of DNA; yet there seemed to be so much of this non-coding DNA that its significance could not be ignored. Non-coding DNA makes up a significant portion of the total genomic DNA in many eukaryotes. For example, older sources estimate 97% of the human genome to be non-coding DNA, while the recently published sequence data increases the estimates to 98.9% non-coding DNA. Introns, the DNA sequences that interrupt coding sequences and do not code for proteins themselves along with other non-coding DNA, play an important role in repression of genes and the sequential switching of genes during development, suggesting that up to 15% of "junk DNA" functions in this vital role (Standish, 2002).

## 2.4 Long-Range Correlations in DNA Base Sequence

DNA sequences represent a condensed archive of information on the structure and function of DNA, both as a complex machinery inside the cell and as the genetic memory for the entire organism. DNA may be taken as representative of the simultaneous needs of order and plasticity of living systems. Therefore, the characterization of the relation between DNA structure and function and the statistical properties of the distribution of its nucleotides may offer us a reliable basis for the further development of a holistic approach. The statistical properties of DNA sequences have been studied extensively in the last 15 years. The general result that emerges from these studies is that DNA statistics is characterized by short-range and long-range correlations which are linked to the functional role of the sequences. Specifically, while coding sequences seem to be almost uncorrelated, non-coding sequences show long-range power-law correlations typical of scale invariant systems (Buiatti and Buiatti, 2004).

DNA topology is of fundamental importance for a wide range of biological processes (Bates and Maxwell, 1993). One big question in DNA research is whether there is some meaning to the order of the base pairs in DNA. Human DNA has become a fascinating topic for physicists to study. One reason for this fascination is the fact that when living cells divide, the DNA is replicated exactly. This is interesting because approximately 95% of human DNA is called "junk" even by biologists who specialize in DNA. One practical task for physicists is simply to identify which sequences within the molecule are the coding sequences. Another scientific interest is to discover why the "junk" DNA is there in the first place. Almost everything in biology has a purpose that, in principle, is discoverable (Stanley, 2000). The study of statistical patterns in DNA sequences is important as it may improve our understanding of the organization and evolution of life on the genomic level. Recent studies indicate that the DNA sequence of letters A, C, G and T does have a $1/f^{\alpha}$ frequency spectrum where $f$ is the frequency and $\alpha$ the exponent. It is possible, therefore, that the sequences have long-range order and underlying grammar rules. The opinion on this issue remains divided (Som *et al.*, 2001 and all references therein). The findings of long-range correlations (LRC) in DNA sequences have attracted much attention, and attempts have been made to relate those findings to known biological features such as the presence of triplet periodicities in protein-coding DNA sequences, the evolution of DNA sequences, the length distribution of protein-coding regions, or the expansion of simple sequence repeats (Holste *et al.*, 2001).

A summary of recent results relating to LRC in DNA sequences is given in the following. Based on spectral analyses, Li *et al.* found (Li, 1992; Li and Kaneko, 1992; Li *et al.*, 1994) that the frequency spectrum of a DNA sequence containing mostly *introns* shows $1/f^{\alpha}$ behavior, which evidence the presence of LRC. The correlation properties of coding and non-coding DNA sequences were first studied by Peng *et al.* (1992) in their *fractal* landscape or DNA walk model. Peng *et al.* (1992) discovered that there exists LRC in non-coding DNA sequences while the coding sequences correspond to a regular random walk. By doing a more detailed analysis of the same data set, Chatzidimitriou-Dreismann and Larhammar (1993) concluded that both coding and non-coding sequences exhibit LRC. A subsequent work by Prabhu and Claverie (1992) also substantially corroborates these results. Buldyrev *et al.* (1995) showed the LRC appears mainly in non-coding DNA using all the DNA sequences available. Alternatively, Voss (1992, 1994), based on equal-symbol correlation, showed a power-law behavior for the sequences studied regardless of the percent of *intron* contents. Havlin *et al.* (1995) state that DNA sequence in genes containing non-coding regions is correlated, and that the correlation is remarkably long range indeed, thousands of distant base pairs are correlated. Such LRC are not found in the coding regions of the gene. Havlin *et al.* (1995) suggest that non-coding regions in plants and invertebrates may display a smaller entropy and larger redundancy than coding regions, further supporting the possibility that non-coding regions of DNA may carry biological information. Investigations based on different models seem to suggest different results, as they all look into only a certain aspect of the entire DNA sequence. It is therefore important to investigate the degree of correlations in a model-independent way. Hence, one may ignore the composition of the four kinds of bases in coding and non-coding segments and only consider the rough structure of the complete genome or long DNA sequences. Yu *et al.* (2000) proposed a time series model based on the global structure of the complete genome and considered three kinds of length sequences. The values of the exponents from these three kinds of length sequences of bacteria indicate that the LRC exist in most of these sequences (Yu *et al.*, 2000 and all the references contained therein). Recently, from a systematic analysis of human *exons*, coding sequences (CDS) and *introns*, Audit *et al.* (2001) have found that power law correlations (PLC) are not only present in non-coding sequences but also in coding regions somehow hidden in their inner codon structure. If it is now well admitted that LRC do exist in genomic sequence, their biological interpretation is still a continuing debate (Audit *et al.*, 2001 and all references therein).

The LRC does not necessarily imply a deviation from Gaussianity. For example, the fractional Brownian motion, which has Gaussian statistics, shows an inverse power-law spectrum. According to Allegrini *et al.* (1996, based on Levy's statistics), LRC would imply a strong deviation from Gaussian statistics while the investigation of Arneodo *et al.* (1995) yields an important conclusion that the DNA statistics are essentially Gaussian (Mohanty and Narayana Rao, 2000).

In visualizing very long DNA sequences, including the complete genomes of several bacteria, yeast and segments of human genes, it is seen that *fractal*-like patterns underlie these biological objects of prominent importance. The method used to visualize genomes of organisms may well be used as a convenient tool to trace, e.g., evolutionary relatedness of species (Hao *et al.*, 2000). Stanley *et al.* (1996) and Stanley (1996a) discuss examples of complex systems composed of many interacting subsystems, which display non-trivial LRC or long-term "memory". The statistical properties of DNA sequences, heartbeat intervals, brain plaque in Alzheimer brains, and fluctuations in economics have the common feature that the guiding principle of scale invariance and universality appear to be relevant (Stanley, 2000).

## 2.5  Emergence of Order and Coherence in Biology

The problem of emergence of macroscopic variables out of microscopic dynamics is of crucial relevance in biology (Vitiello, 1992). Biological systems rely on a combination of network and the specific elements involved (Kitano, 2002). The notion that membership in a network could confer stability emerged from Ludwig von Bertalanffy's description of general systems theory in the 1930s and Norbert Wieners description of cybernetics in the 1940s. General systems theory focused in part on the notion of flow, postulating the existence and significance of flow equilibria. In contrast to Cannon's concept that mechanisms should yield homeostasis, general systems theory invited biologists to consider an alternative model of homeodynamics in which nonlinear, non-equilibrium processes could provide stability, if not constancy (Buchman, 2002).

The cell dynamical system model for coherent pattern formation in turbulent flows summarized earlier (Chapter 1) may provide a general systems theory for biological complexity. General systems theory is a logical-mathematical field, the subject matter of which is the formulation and deduction of those principles which are valid for "systems" in general, whatever be the nature of their component elements or the relations or "forces" between them (Von Bertalanffy, 1968; Peacocke, 1989; Klir, 1992).

## 2.6  Multidisciplinary Approach for Modelling Biological Complexity

Computational biology involves extraction of the hidden patterns from huge quantities of experimental data, forming hypotheses as a result, and simulation-based analyses, which tests hypotheses with experiments, providing predictions to be tested by further studies. Robust systems maintain their state and functions against external and internal perturbations, and robustness is an essential feature of biological systems. Structurally stable network configurations in biological systems increase insensitivity to parameter changes, noise and minor mutations (Kitano, 2002). Systems biology advocates a multidisciplinary approach for modelling biological complexity. Many features of biological complexity result from self-organization. Biological systems are, in general, global patterns produced by local interactions. One of the appealing aspects of the study of self-organized systems is that we do not need anything specific from biology to understand the existence of self-organization. Self-organization occurs for reasons that have to do with the organization of the interacting elements (Cole, 2002). The first and most general criterion for systems thinking is the shift from the parts to the whole. Living systems are integrated wholes whose properties cannot be reduced to those of smaller parts (Capra, 1996). Many disciplines may have helpful insights to offer or useful techniques to apply to a given problem, and to the extent that problem-focused research can bring together practitioners of different disciplines to work on shared problems, this can only be a good thing. A highly investigated but poorly understood phenomena, is the ubiquity of the so-called $1/f$ spectra in many interesting phenomena, including biological systems (Wooley and Lin, 2005).

## 2.7 Fractal Fluctuations and Statistical Normal Distribution

Statistical and mathematical tools are used for analysis of data sets and estimation of the probabilities of occurrence of events of different magnitudes in all branches of science and other areas of human interest. Historically, the statistical normal or the Gaussian distribution has been in use for nearly 400 years and gives a good estimate for probability of occurrence of the more frequent moderate-sized events of magnitudes within two standard deviations from the mean. The Gaussian distribution is based on the concept of data independence, fixed mean and standard deviation with a majority of data events clustering around the mean.

However, for real world infrequent hazardous extreme events of magnitudes greater than two standard deviations, the statistical normal distribution gives progressively increasing under-estimates of up to near zero probability. In the 1890s, the power law or Pareto distributions with implicit LRC were found to fit the fat tails exhibited by hazardous extreme events such as heavy rainfall, stock market crashes, traffic jams, the after-shocks following major earthquakes, etc. A historical review of statistical normal and the Pareto distributions are given by Andriani and McKelvey (2007) and Selvam (2009). The spatial and/or temporal data sets in practice refer to real world or computed dynamical systems and are fractals with self-similar geometry and LRC in space and/or time, i.e., the statistical properties such as the mean and variance are scale-dependent and do not possess fixed mean and variance and therefore the statistical normal distribution cannot be used to quantify/describe self-similar data sets. Though the observed power law distributions exhibit qualitative universal shape, the exact physical mechanism underlying such scale-free power laws is not yet identified for the formulation of universal quantitative equations for fractal fluctuations of all scales. The universal inverse power law for fractal fluctuations is shown to be a function of the golden mean based on general systems theory concepts for fractal fluctuations (Chapter 1).

## Acknowledgement

The author is grateful to Dr. A. S. R. Murty for encouragement.

## References

Alcamo, E. (2001). *DNA Technology*, 2nd Ed. (Academic Press, New York), p. 339.

Allegrini, P., Barbi, M., Grigolini, P., and West, B. J. (1996). Dynamical model for DNA sequences, *Phys. Rev E*, 52(5), pp. 5281–5296. http://linkage.rockefeller.edu/wli/dna_corr.

Andriani, P. and McKelvey, B. (2007). Beyond Gaussian averages: Redirecting management research toward extreme events and power laws, *J. Int. Bus. Stud.*, 38, pp. 1212–1230.

Arber, A. (1950) *The Natural Philosophy of Plant Form* (Cambridge University Press, London).

Arneodo, A., Bacry, E., Graves, P. V. and Muzy, J. F. (1995). Characterizing long-range correlations in DNA sequences from wavelet analysis, *Phys. Rev. Lett.*, 74(16), pp. 3293–3296. http://linkage.rockefeller.edu/wli/dna_corr/arneodo95.pdf.

Arneodo, A., Muzy, J. -F., and Sornette, D. (1998). "Direct" causal cascade in the stock market, *Eur. Phys. J. B*, 2, pp. 277–282.

Audit, B., Thermes, C., Vaillant, C., d'Aubenton-Carafa, Y., Muzy, J. F., and Arneodo, A. (2001). Long-range correlations in genomic DNA: A signature of the nucleosomal structure, *Phys. Rev. Lett.*, 86(11), pp. 2471–2474. http://linkage.rockefeller.edu/wli/dna_corr/audit01.pdf.

Bak P. C. and Chen, K. (1989). The physics of fractals, *Physica D*, 38(1–3), pp. 5–12.

Bak P. C. and Chen, K. (1991). Self-organized criticality, *Sci. Amer.*, 264, pp. 46–53.

Bak, P., Tang, C., and Wiesenfeld, K. (1987). Self-organized criticality: An explanation of 1/f noise, *Phys. Rev. Lett.*, 59, pp. 381–384.

Bak P. C., Tang C., and Wiesenfeld, K. (1988). Self-organized criticality, *Phys. Rev. Ser. A*, 38, pp. 364–374.

Bak, P., Chen, K., Scheinkman, J. A., and Woodford, M. (1992). *Self-organized Criticality and Fluctuations in Economics* (Sante Fe Institute, Santa Fe). http://www.santafe.edu/sfi/publications/Abstracts/92-04-018abs.html

Ball, P. (2000). Augmenting the alphabet, *Nat. Sci. Update* 30 August.

Bates, A. D. and Maxwell, A. (1993). *DNA Topology* (Oxford University Press, Oxford).

Bernaola-Galvan, P., Carpena, P., Roman-Roldan, R., and Oliver, J. L. (2002). Study of statistical correlations in DNA sequences, *Gene*, 300(1–2), pp. 105–115. http://www.nslij-genetics.org/dnacorr/bernaola02.pdf.

Bernardi, G., Olofsson, B., Filipski, J., Zerial, M., Salinas, J., Cuny, G., Meunier-Rotival, M., and Rodier, F. (1985). The mosaic genome of warm-blooded vertebrates, *Science*, 228, pp. 953–958.

Bernardi, G. (1989). The isochore organization of the human genome, *Annu. Rev. Genet.*, 23, pp. 637–661.

Buchanan, M. (1997). One law to rule them all, *New Scientist*, 8 November, pp. 30–35.

Buchman, T. G. (2002). The community of the self, *Nature*, 420, pp. 246–251.

Buiatti, M. and Buiatti, M. (2004). Towards a statistical characterization of the living state of matter, *Chaos Solit. Fractals*, 20, pp. 55–61.

Buldyrev, S. V., Goldberger, A. L., Havlin, S., Mantegna, R. N., Matsa, M. E., Peng, C. K., Simons, M., and Stanley, H. E. (1995). Long-range correlation properties of coding and non-coding DNA sequences — GenBank analysis, *Phys. Rev. E*, 51(5), pp. 5084–5091. http://linkage.rockefeller.edu/wli/dna_corr/buldyrev95.pdf.

Capra, F. (1996). *The Web of Life* (Harper Collins, London).

Chatzidimitriou-Dreismann, C. A., and Larhammar, D. (1993). Long-range correlations in DNA, *Nature*, 361, pp. 212–213.

Check, E. (2006). It's the junk that makes us human, *Nature*, 444, pp. 130–131.

Chen, P. (1996a). Trends, shocks, persistent cycles in evolving economy: Business cycle measurement in time-frequency representation. In *Nonlinear Dynamics and Economics*, eds. Barnett, W. A., Kirman, A. P., and Salmon, M., Chapter 13, pp. 307–331 (Cambridge University Press, Cambridge).

Chen, P. (1996b). A random walk or color chaos on the stock market? Time-frequency analysis of S&P indexes, *Stud. Nonlinear Dyn. Econom.*, 1(2), pp. 87–103.

Claire, J. (1964). *The Stuff of Life* (Phoenix House, London), p. 67.

Clark, A. G. (2001). The search for meaning in noncoding DNA, *Genome Res.*, 11, pp. 1319–1320. http://linkage.rockefeller.edu/wli/dna_corr.

Cole, B. J. (2002). Evolution of self-organized systems, *Biol. Bull.*, 202, pp. 256–261.

Thompson, D. W. (1963). *On Growth and Form*, 2nd Ed. (Cambridge University Press, Cambridge).

Deering, W., West, B. J. (1992). Fractal physiology, *IEEE Eng. Med. Biol.*, June, pp. 40–46.

Eatwell, J., Milgate, M., and Newman, P. (1991). *The New Palgrave: A Dictionary of Economics*, Vol. 3 (MacMillan Press, London).

Farmer, J. D. (1999). Physicists attempt to scale the ivory towers of finance, *Computing in Science & Engineering*, November/December, pp. 26–39. http://www.santafe.edu/sfi/publications/Abstracts/99-10-073abs.html.

Feigenbaum, J. A. and Freund, P. G. O. (1997a). Discrete scaling in stock markets before crashes. http://xxx.lanl.gov/pdf/cond-mat/9509033.

Feigenbaum, J. A. and Freund, P. G. O. (1997b). Discrete scale invariance and the second Black Monday. http://xxx.lanl.gov/pdf/cond-mat/9710324.

Feigenbaum, J. A. (2001a). A statistical analysis of log-periodic precursors to financial crashes. http://xxx.lanl.gov/pdf/cond-mat/0101031.

Feigenbaum, J. A. (2001b). More on a statistical analysis of log-periodic precursors to financial crashes. http://xxx.lanl.gov/pdf/cond-mat/0107445.

Freeman, G. R. (1987). Introduction. In *Kinetics of Nonhomogenous Processes*, ed. Freeman, G. R. , pp. 1–18 (John Wiley and Sons, Inc., New York).

Freeman, G. R. (1990). *KNP*89: Kinetics of non-homogenous processes (KNP) and nonlinear dynamics, *Can. J. Phys.*, 68, pp. 655–659.

Ghashghaie, S., Wolfgang, B., Joachim, P., Peter, T., and Yadolah, D. (1996). Turbulent cascades in foreign exchange markets, *Nature*, 381(6585), pp. 767–770.

Gleick, J. (1987). *Chaos: Making a New Science* (Viking, New York).

Goldberger, A. L., Amaral, L. A. N., Hausdorff, J. M., Ivanov, P. Ch., Peng, C.-K., and Stanley, H. E. (2002). Fractal dynamics in physiology: Alterations with disease and aging, *PNAS*, 99, Suppl. 1, pp. 2466–2472. http://pnas.org/cgi/doi/10.1073/pnas.012579499.

Goldberger, A. L., Rigney, D. R., and West, B. J. (1990). Chaos and fractals in human physiology, *Sci. Am.*, 262(2), pp. 42–49.

Gopikrishnan, P., Plerou, V., Amaral, L., Meyer, M., and Stanley, H. E. (1999). Scaling of the distribution of fluctuations of financial market indices, *Phys. Rev. E*, LX, pp. 5305–5316.

Gribbin, J. (1985) *In Search of the Double Helix* (Wildwood House Ltd., England), p. 362.

Guharay, S., Hunt, B. R., Yorke, J. A., and White, O. R. (2000). Correlations in DNA sequences across the three domains of life, *Physica D*, 146, pp. 388–396. http://linkage.rockefeller.edu/wli/dna_corr/guharay00.pdf.

Gutenberg, R. and Richter, C. F. (1944). Frequency of earthquakes in California, *Bull. Seismol. Soc. Am.*, 34, pp. 185–188.

Hooge, C., Lovejoy, S., Schertzer, D., Pecknold, S., Malouin, J. F., and Schmitt, F. (1994). Multifractal phase transitions: The origin of self-organized criticality in earthquakes, *Nonlinear Processes Geophys.*, 1, pp.191–197.

Hao Bailin, Lee, H., and Zhang, S. (2000). Fractals related to long DNA sequences and complete genomes, *Chaos Solit. Fractal.*, 11(6), pp. 825–836. http://linkage.rockefeller.edu/wli/dna_corr/haolee00.pdf.

Havlin S., Buldyrev S. V., Goldberger, A. L., Mantegna, R. N., Peng, C. K., Simons, M., Stanley, H. E. (1995). Statistical and linguistic features of DNA sequences, *Fractals*, 221, pp. 269–284.

Holste, D., Grosse, I., and Herzel, H. (2001). Statistical analysis of the DNA sequence of human chromosome 22, *Phys. Review E*, 64, pp. 041917(1–9). http://linkage.rockefeller.edu/wli/dna_corr/holste01.pdf.

Ikemura, T. (1985). Codon usage and tRNA content in unicellular and multicellular organisms, *Mol. Biol. Evol.*, 2, pp. 13–34.

Ikemura, T. and Aota, S. (1988). Global variation in G + C content along verte-brate genome DNA: Possible correlation with chromosome band structures, *J. Mol. Biol.*, 203, pp. 1–13.

Jean, R. V. (1994). *Phyllotaxis: A Systemic Study in Plant Morphogenesis* (Cambridge University Press, New York).

Kadanoff, L. P. (1996). Turbulent excursions, *Nature*, 382, pp. 116–117.

Kettlewell, J. (2004). "Junk" throws up precious secret, *BBC News*, 12 May. http://news.bbc.co.uk.

Kitano, H. (2002). Computational systems biology, *Nature*, 420, pp. 206–210.

Klir, G. J. (1992). Systems science: A guided tour, *J. Biol. Sys.*, 1, pp. 27–58.

Kolmogorov, A. N. (1941). The local structure of turbulence in incompressible liquids for very high Reynolds numbers, *C. R. Russ. Acad. Sci.*, 30, pp. 301–305.

Kolmogorov, A. N. (1962). A refinement of previous hypotheses concerning the local structure of turbulence in a viscous inhomogeneous fluid at high Reynolds number, *J. Fluid Mech.*, 13, pp. 82–85.

Leone, F. (1992). *Genetics: The Mystery and the Promise* (TAB Books, McGraw Hill, Inc.), p. 229.

Li, W. and Holste, D. (2004). Spectral analysis of guanine and cytosine concen-tration of mouse genomic DNA, *Fluct. Noise Lett.*, 4(3), pp. L453–L464. http://www.nslij-genetics.org/wli/pub/fnl04.pdf.

Li, W. and Kaneko, K. (1992). Long-range correlation and partial $1/f^{\alpha}$ spectrum in a noncoding DNA sequence, *Europhys. Lett.*, 17(7), pp. 655–660. http://linkage.rockefeller.edu/wli/dna_corr/l-epl92-lk.html.

Li, W. (1992). Generating nontrivial long-range correlations and $1/f$ spectra by replication and mutation, *Int. J. Bifur. Chaos*, 2(1), pp. 137–154. http://link-age.rockefeller.edu/wli/dna_corr/l-ijbc92-l.html.

Li, W., Marr, T. G., and Kaneko, K. (1994). Understanding long-range correla-tions in DNA sequences, *Physica* D, 75(1–3), pp. 392–416 (1994); erratum: 82, p. 217. http://arxiv.org/chao-dyn/9403002.

Lovejoy, S. and Schertzer, D. (1986). Scale invariance, symmetries, fractals and stochastic simulations of atmospheric phenomena, *Bull. Amer. Meteorol. Soc.*, 67, pp. 21–32.

Macdonald, G. F. (1989). Spectral analysis of time series generated by nonlinear processes, *Rev. Geophys.*, 27, pp. 449–469.

Mandelbrot, B. B. (1975). On the geometry of homogenous turbulence with stress on the fractal dimension of the iso-surfaces of scalars, *J. Fluid Mech.*, 72, pp. 401–416.

Mandelbrot, B. B. (1977). *Fractals: Form, Chance and Dimension* (Freeman, San Francisco).

Mantegna, R. N. and Stanley, H. E. (1995). Scaling behaviour in the dynamics of an economic index, *Nature*, 376, pp. 46–49.

Mary Selvam, A. (1990). Deterministic chaos, fractals and quantumlike mechanics in atmospheric flows, *Can. J. Phys.*, 68, pp. 831–841. http://xxx.lanl.gov/html/physics/0010046.

Mary Selvam, A. (1998) Quasicrystalline pattern formation in fluid substrates and phyllotaxis. In *Symmetry in Plants*, eds. Barabe, D. and Jean, R. V., Vol. 4, pp.795–809 (World Scientific Series in Mathematical Biology and Medicine, Singapore). http://xxx.lanl.gov/abs/chao-dyn/9806001.

Mary Selvam, A., Pethkar, J. S. and Kulkarni, M. K. (1992). Signatures of a universal spectrum for atmospheric interannual variability in rainfall time series over the Indian Region, *Int'l J. Climatol.*, 12, pp. 137–152.

Mohanty, A. K. and Narayana Rao, A. V. S. S. (2000). Factorial moments analyses show a characteristic length scale in DNA sequences, *Phys. Rev. Lett.*, 84(8), pp. 1832–1835. http://linkage.rockefeller.edu/wli/dna_corr/mohanty00.pdf.

Monin, A. S. and Yaglom, A. M. (1975). *Statistical Hydrodynamics*, Vols. 1 and 2 (MIT Press, Cambridge).

Muller, A. and Beugholt, C. (1996). The medium is the message, *Nature*, 383, pp. 296–297.

Newman, M. (2000). The power of design, *Nature*, 405, pp. 412–413.

Omori, F. (1895). On the aftershocks of earthquakes, *J. Coll. Sci.*, 7, p. 111.

Peacocke, A. R. (1989). *The Physical Chemistry of Biological Organization* (Clarendon Press, Oxford).

Peng, C.-K., Buldyrev, S. V., Goldberger, A. L., Havlin, S., Sciortino, F., Simons, M., and Stanley, H. E. (1992). Long-range correlations in nucleotide sequences, *Nature*, 356, pp. 168–170.

Plerou, V., Gopikrishnan, P., Amaral, L., Meyer, M., and Stanley, H. E. (1999). Scaling of the distribution of fluctuations of financial market indices, *Phys. Rev. E*, LX, pp. 6519–6529.

Prabhu, V. V. and Claverie, J. M. (1992). Correlations in intronless DNA, Scientific Correspondence, *Nature*, 359, p. 782. http://linkage.rockefeller.edu/wli/dna_corr.

Richardson, L. F. (1960) The problem of contiguity: An appendix to statistics of deadly quarrels. In *General Systems — Year Book of the Society for General Systems Research*, eds. Von Bertalanffy, L. and Rapoport, A., Vol. V, pp. 139–187 (Ann Arbor, Michigan).

Schroeder, M. (1991). *Fractals, Chaos and Power-laws* (W. H. Freeman and Co., New York).

Selvam, A. M. (2009). Fractal fluctuations and statistical normal distribution, *Fractals*, 17(3), pp. 333–349. http://arxiv.org/pdf/0805.3426.

Selvam, A. M. and Fadnavis, S. (1998). Signatures of a universal spectrum for atmospheric interannual variability in some disparate climatic regimes, *Meteorol. & Atmos. Phys.*, 66, pp. 87–112. http://xxx.lanl.gov/abs/chao-dyn/9805028.

Skinner, J. E. (1994). Low dimensional chaos in biological systems, *Bio/technology*, 12, pp. 596–600.

Som, A. Chattopadhyay, S., Chakrabarti, J. and Bandyopadhyay, D. (2001). Codon distributions in DNA, *Phys. Rev. E*, 63(5), 051908, pp. 1–8. http://linkage.rockefeller.edu/wli/dna_corr/som01.pdf.

Sornette, D, Johansen, A., and Bouchaud, J.-P. (1996). Stock market crashes, precursors and replicas. *J. Phys. I*, 6, pp. 167–175. arXiv:cond-mat/9510036v1.

Standish, T. G. (2002). Rushing to judgment: Functionality in noncoding or "junk" DNA, *ORIGINS*, 53, pp. 7–30.

Stanley H. E., Amaral, L. A. N., Gopikrishnan, P., and Plerou, V. (2000). Scale invariance and universality of economic fluctuations, *Physica A*, 283, pp. 31–41.

Stanley, H. E. (1995). Powerlaws and universality, *Nature*, 378, p. 554.

Stanley, H. E. (2000). Exotic statistical physics: Applications to biology, medicine, and economics, *Physica A*, 285, pp. 1–17.

Stanley, H. E., Afanasyev, V., Amaral, L. A. N., Buldyrev, S. V., Goldberger, A. L., Havlin, S., Leschhorn, H., Maass, P., Mantegna, R. N., Peng, C.-K., Prince, P. A., Salinger, M. A., Stanley, M. H. R., and Viswanathanan, G. M. (1996a). Anomalous fluctuations in the dynamics of complex systems: From DNA and physiology to econophysics, *Physica A*, 224(1–2), pp. 302–321.

Stanley, H. E., Amaral, L. A. N., Buldyrev, S. V., Goldberger, A. L., Havlin, S., Hyman, B. T., Leschhorn, H., Maass, P., Makse, H. A., Peng, C.-K., Salinger, M. A., Stanley, M. H. R., and Vishwanathan, G. M. (1996b). Scaling and universality in living systems, *Fractals*, 4(3), pp. 427–451.

Stanley, M. H. R., Amaral, L. A. N., Buldyrev, S. V., Havlin, S., Leschhorn, H. Maass, P., Salinger, M. A., and Stanley H. E. (1996c). Can statistical physics contribute to the science of economics? *Fractals*, 4(3), pp. 415–425.

Stewart, I. (1992). Where do nature's patterns come from? *New Scientist*, 135, p. 14.

Stewart, I. (1995). Daisy, daisy, give your answer do, *Sci. Amer.*, 272, pp. 76–79.

Strathmann, R. R. (1990). Testing size abundance rules in a human exclusion experiment, *Science*, 250, p. 1091.

Taft, R. J. and Mattick, J. S. (2003). Increasing biological complexity is positively correlated with the relative genome-wide expansion of non-protein-coding DNA sequences, *Genome Biol.*, 5, p. 1. http://genomebiology.com/2003/5/1/P1.

Thompson, D. W. (1963). *On Growth and Form*, 2nd Ed. (Cambridge University Press, Cambridge).

Vitiello, G. (1992). Coherence and electromagnetic fields in living matter, *Nanobiology*, 1, pp. 221–228.

Von Bertalanffy, L. (1968). *General Systems Theory: Foundations, Development, Applications* (George Braziller, New York).

Voss, R. F. (1992). Evolution of long-range fractal correlations and $1/f$ noise in DNA base sequences, *Phys. Rev. Lett.*, 68(25), pp. 3805–3808.

Voss, R. F. (1994). Long-range fractal correlations in DNA introns and exons, *Fractals*, 2(1), pp. 1–6.

Watson, J. D. and Crick, F. H. C. (1953). A structure for deoxyribose nucleic acid, *Nature*, April 25, pp. 737–738.

Watson, J. D. (1997) *The Double Helix* (Weidenfeld and Nicolson, London), p.175.

West, B. J. (2004). Comments on the renormalization group, scaling and measures of complexity, *Chaos Solit. Fractal*, 20, pp. 33–44.

West, B. J. (1990a). Fractal forms in physiology, *Int'l. J. Mod. Phys. B*, 4(10), pp. 1629–1669.

West, B. J. (1990b). Physiology in fractal dimensions, *Ann. Biomed. Eng.*, 18, pp. 135–149.

Wooley, J. C. and Lin, H. S. (2005). Illustrative problem domains at the interface of computing and biology. In *Catalyzing Inquiry at the Interface of Computing and Biology*, eds. Wooley, J. C. and Lin, H. S. (Computer Science and Telecommunications Board, USA: The National Academies Press), p. 329.

Yu, Z-G., Anh, V. V. and Wang, B. (2000). Correlation property of length sequences based on global structure of the complete genome, *Phys. Rev. E*, 63, pp. 011903(1–8). http://linkage.rockefeller.edu/wli/dna_corr/yu00.pdf.

# Chapter 3

# Long-Range Correlations Data 1: Universal Spectrum for DNA Base C+G Frequency Distribution in Human Chromosomes 1–24*

## 3.1 Introduction

Spatially extended dynamical systems in nature exhibit fractal space-time fluctuations associated with inverse power law spectrum or $1/f$ noise, signifying long-range space-time correlations or memory, identified as self-organized criticality (Bak, 1988; Goldberger *et al.*, 2002; West, 2004; Milotti, 2004; Wooley and Herbert, 2005; Wang *et al.*, 2006; Li, 2007). Self-similar space-time fluctuations, now identified as fractals, have earlier been reported and studied in separate branches of science till very recently (1980s) when they were recognized as universal phenomena under the new science of "nonlinear dynamics and chaos". The multidisciplinary nature of investigations will help gain new insights and develop mathematical and statistical techniques and analytical tools for understanding and quantifying the physics of the observed long-range correlations in dynamical systems in nature. The physics of dynamical systems therefore comes under the broad category of general systems theory. The

* http://arxiv.org/pdf/physics/0701079 (2009). *World J. Model. Simul.*, Vol. 5(2), pp. 151–160.

sub-units of the system function as a unified whole two-way communication and control network with global (system level) control/response to local functions/stimuli, thereby possessing the criteria for a robust system (Csete and Doyle, 2002; Kitano, 2002, 2004). Kitano (2002) makes the point that robustness is a property of an entire system; it may be that no individual component or process within a system would be robust, but the system-wide architecture still provides robust behaviour. This presents a challenge for analysis, since elucidating such behaviours can be counter-intuitive and computationally demanding.

DNA sequences represent a condensed archive of information on the structure and function of DNA, both as a complex machinery inside the cell and as the genetic memory for the entire organism. DNA may be taken as representative of the simultaneous needs of order and plasticity of living systems. Therefore, the characterization of the relation between DNA structure and function and the statistical properties of the distribution of its nucleotides may offer us a reliable basis for the further development of a holistic approach. The statistical properties of DNA sequences have been studied extensively in the last 15 years. The general result that emerges from these studies is that DNA statistics is characterized by short-range and long-range correlations which are linked to the functional role of the sequences. Specifically, while coding sequences seem to be almost uncorrelated, non-coding sequences show long-range power-law correlations typical of scale invariant systems (Buiatti and Buiatti, 2004).

During the late 1960s papers began appearing that showed eukaryotic DNA contained large quantities of repetitive DNA which did not appear to code for proteins. By the early 1970s, the term "junk DNA" had been coined to refer to this non-coding DNA. Junk DNA seemed like an appropriate term for DNA cluttering up the genome while contributing in no way to the protein coding function of DNA; yet there seemed to be so much of this non-coding DNA that its significance could not be ignored. Non-coding DNA makes up a significant portion of the total genomic DNA in many eukaryotes. For example, older sources estimate 97% of the human genome to be non-coding DNA, while the recently published sequence data increases the estimates to 98.9% non-coding DNA. Introns, the DNA sequences that interrupt coding sequences and do not code for proteins themselves along with other non-coding DNA, play an important role in repression of genes and the sequential switching of genes during development, suggesting that up to 15% of "junk DNA" functions in this vital role (Standish, 2002).

In this chapter, it is shown that the spectra of human chromosomes 1–24 DNA base C+G frequency distributions follow the universal inverse power law form of the statistical normal distribution consistent with predictions of a recently developed general systems theory model for dynamical systems of all space-time scales.

## 3.2 General Systems Theory Concepts

In summary (Selvam, 1990; Selvam and Fadnavis, 1998), the model is based on Townsend's concept (Townsend, 1956) that large eddy structures form in turbulent flows as envelopes of enclosed turbulent eddies. Such a simple concept that space-time averaging of small-scale structures gives rise to large-scale space-time fluctuations leads to the following important model predictions.

### 3.2.1 *Quantum-like chaos in turbulent fluid flows*

Since the large eddy is but the integrated mean of enclosed turbulent eddies, the eddy energy (kinetic) distribution follows statistical normal distribution according to the Central Limit Theorem (Ruhla, 1992). Such a result, that the additive amplitudes of the eddies, when squared, represent probability distributions is found in the subatomic dynamics of quantum systems such as the electron or photon. Atmospheric flows, or, in general turbulent fluid flows follow quantum-like chaos.

### 3.2.2 *Dynamic memory (information) circulation network*

The root mean square (r.m.s.) circulation speeds $W$ and $w_*$ of large and turbulent eddies of respective radii $R$ and $r$ are related as

$$W^2 = \frac{2}{\pi}\frac{r}{R}w_*^2 \qquad (3.1)$$

Equation (3.1) is a statement of the law of conservation of energy for eddy growth in fluid flows and implies a two-way ordered energy flow between the larger and smaller scales. Microscopic scale perturbations are carried permanently as internal circulations of progressively larger eddies. Fluid flows therefore act as dynamic memory circulation networks with intrinsic long-term memory of short-term fluctuations.

### 3.2.3 *Quasicrystalline structure*

The flow structure consists of an overall logarithmic spiral trajectory with Fibonacci winding number and quasiperiodic Penrose tiling pattern for internal structure (Fig. 3.1). Primary perturbation $OR_O$ (Fig. 3.1) of time period $T$ generates return circulation $OR_1R_O$ which, in turn, generates

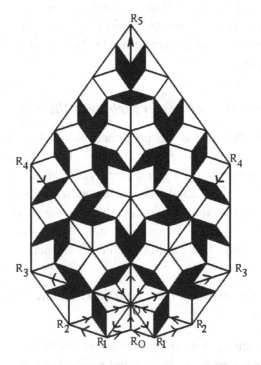

**Figure 3.1:**   Internal structure of large eddy circulations. Large eddies trace an overall logarithmic spiral trajectory $OR_OR_1R_2R_3R_4R_5$ simultaneously in clockwise and anti-clockwise directions with the quasiperiodic Penrose tiling pattern for the internal structure.

successively larger circulations $OR_1R_2$, $OR_2R_3$, $OR_3R_4$, $OR_4R_5$, etc., such that the successive radii form the Fibonacci mathematical number series, i.e., $OR_1/OR_0 = OR_2/OR_1 = \ldots\ldots = \tau$ where $\tau$ is the golden mean equal to $(1+\sqrt{5})/2 \approx 1.618$. The flow structure therefore consists of a nested continuum of vortices, i.e., vortices within vortices.

The quasiperiodic Penrose tiling pattern with five-fold symmetry has been identified as quasicrystalline structure in condensed matter physics (Janssen, 1988). The self-organized large eddy growth dynamics, therefore, spontaneously generates an internal structure with the five-fold symmetry of the dodecahedron, which is referred to as the icosahedral symmetry. Recently the carbon macromolecule $C_{60}$, formed by condensation from a carbon vapour jet, was found to exhibit the icosahedral symmetry of the closed soccer ball and has been named Buckminsterfullerene or footballene (Curl and Smalley, 1991). Self-organized quasicrystalline pattern formation therefore exists at the molecular level also and may result in condensation of specific biochemical structures in biological media. Logarithmic spiral formation with Fibonacci winding number and five-fold symmetry possess maximum packing efficiency for component parts and are manifested strikingly in plant *Phyllotaxis* (Jean, 1994).

### 3.2.4 *Dominant periodicities*

Dominant quasiperiodicities $P_n$ corresponding to the internal circulations (Fig. 3.1) $OR_0R_1$, $OR_1R_2$, $OR_2R_3$, ... are given as

$$P_n = T(2 + \tau)\tau^n \qquad (3.2)$$

The dominant quasiperiodicities are equal to $2.2T$, $3.6T$, $5.8T$, $9.5T$, ... for values of $n = -1, 0, 1, 2,...$, respectively (Eq. (3.2)). Space-time integration of turbulent fluctuations results in robust broadband dominant periodicities which are functions of the primary perturbation time period $T$ alone and are independent of exact details (chemical, electrical, physical, etc.) of turbulent fluctuations. Persistent periodic energy pumping at fixed time intervals (period) $T$ in a fluid medium generates self-sustaining continuum of eddies and results in apparent nonlinear chaotic fluctuations in the fluid medium. Also, such global scale oscillations in the unified network are not affected appreciably by failure of localized microscale circulation networks.

Wavelengths (or periodicities) close to the model predicted values have been reported in weather and climate variability (Selvam and Fadnavis, 1998), prime number distribution (Selvam, 2001a), Riemann zeta zeros (non-trivial) distribution (Selvam, 2001b), Drosophila DNA base sequence (Selvam, 2002), stock market economics (Selvam, 2003), Human chromosome 1 DNA base sequence (Selvam, 2004).

Macroscale coherent structures emerge by space-time integration of microscopic domain fluctuations in fluid flows. Such a concept of the autonomous growth of atmospheric eddy continuum with ordered energy flow between the scales is analogous to Prigogine's (Prigogine and Stengers, 1988) concept of the spontaneous emergence of order and organization out of apparent disorder and chaos through a process of self-organization.

### 3.2.4.1 *Emergence of order and coherence in biology*

The problem of emergence of macroscopic variables out of microscopic dynamics is of crucial relevance in biology (Vitiello, 1992). Biological systems rely on a combination of network and the specific elements involved (Kitano, 2002). The notion that membership in a network could confer stability emerged from Ludwig von Bertalanffy's description of general systems theory in the 1930s and Norbert Wieners description of cybernetics in the 1940s. General systems theory focused in part on the notion of flow, postulating the existence and significance of flow equilibria. In contrast to Cannon's concept that mechanisms should yield homeostasis, general systems theory invited biologists to consider an alternative model of homeodynamics in which nonlinear, non-equilibrium processes could provide stability, if not constancy (Buchman, 2002).

The cell dynamical system model for coherent pattern formation in turbulent flows summarized earlier (Chapter 2) may provide a general systems theory for biological complexity. General systems theory is a logical-mathematical field, the subject matter of which is the formulation and deduction of those principles which are valid for "systems" in general, whatever be the nature of their component elements or the relations or "forces" between them (Bertalanffy, 1968; Peacocke, 1989; Klir, 1993).

### 3.2.5 *Long-range spatiotemporal correlations (coherence)*

The logarithmic spiral flow pattern enclosing the vortices $OR_0R_1$, $OR_1R_2$, ... may be visualized as a continuous smooth rotation of the phase angle $\theta$ ($R_0OR_1$, $R_0OR_2$, ... etc.) with increase in period. The phase angle $\theta$ for each stage of growth is equal to $r/R$ and is proportional to the variance $W^2$ (Eq. (3.1)), the variance representing the intensity of fluctuations. The phase angle gives a measure of coherence or correlation in space-time fluctuations. The model predicted continuous smooth rotation of phase angle with increase in period length associated with logarithmic spiral flow structure is analogous to Berry's phase (Berry, 1988; Kepler *et al.*, 1991) in quantum systems.

### 3.2.6 *Universal spectrum of fluctuations*

Conventional power spectral analysis will resolve such a logarithmic spiral flow trajectory as a continuum of eddies (broadband spectrum) with a progressive increase in phase angle. The power spectrum, plotted on log–log scale as variance versus frequency (period) will represent the probability density corresponding to normalized standard deviation $t$ given by

$$t = \frac{\log L}{\log T_{50}} - 1 \tag{3.3}$$

In the aforementioned Eq. (3.3) $L$ is the period in years and $T_{50}$ is the period up to which the cumulative percentage contribution to total variance is equal to 50. The above expression for normalized standard deviation $t$ follows from model prediction of logarithmic spiral flow structure and model concept of successive growth structures by space-time averaging.

The period (or length scale) $T_{50}$ up to which the cumulative percentage contribution to total variances is equal to 50 is computed from model concepts as follows

$$T_{50} = (2 + \tau)\tau^0 \tag{3.4}$$

Fluctuations of all scales therefore self-organize to form the universal inverse power law form of the statistical normal distribution. Since the phase angle $\theta$ equal to $r/R$ represents the variance $W^2$ (Eq. (3.1)), the phase spectrum plotted similar to variance spectrum will also follow the statistical normal distribution.

### 3.2.7   *Quantum mechanics for subatomic dynamics: Apparent paradoxes*

The following apparent paradoxes found in the subatomic dynamics of quantum systems (Maddox, 1988) are consistent in the context of atmospheric flows as explained in the following.

#### 3.2.7.1   *Wave-particle duality*

A quantum system behaves as a wave on some occasions and as a particle at other times. Wave-particle duality is consistent in the context of atmospheric waves, which generate particle-like clouds in a row because of formation of clouds in updrafts and dissipation of clouds in adjacent downdrafts characterizing wave motion (Fig. 3.2).

#### 3.2.7.2   *Non-local connection*

The separated parts of a quantum system respond as a unified whole to local perturbations. Non-local connection is implicit to atmospheric flow structure quantified in Eq. (3.1) as ordered two-way energy flow between larger and smaller scales and seen as long-range space-time correlations, namely self-organized criticality. Atmospheric flows self-organize to form a unified network with the quasiperiodic Penrose tiling pattern for internal

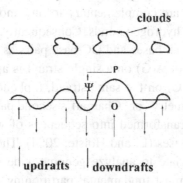

**wave-particle duality**

clouds

Ψ: wave amplitude
P: wave peak
O: observer

**wave-trains in atmospheric flows
and cloud formation**

**Figure 3.2:** Illustration of wave-particle duality as physically consistent for quantum-like mechanics in atmospheric flows. Particle-like clouds form in a row because of condensation of water vapour in updrafts and evaporation of condensed water in adjacent downdrafts associated with eddy circulations in atmospheric flows. Wave-particle duality in macroscale real world dynamical systems may be associated with bimodal (formation and dissipation) phenomenological form for manifestation of energy associated with bidirectional energy flow intrinsic to eddy (wave) circulations in the medium of propagation.

structure (Fig. 3.1), which provide long-range (non-local) space-time connections.

## 3.3 Applications of the General Systems Theory Concepts to Genomic DNA Base Sequence Structure

DNA sequences, the blueprint of all essential genetic information, are polymers consisting of two complementary strands of four types of bases: adenine (A), cytosine (C), guanine (G) and thymine (T). Among the four bases, the presence of A on one strand is always paired with T on the opposite strand, forming a "base pair" with two hydrogen bonds. Similarly, G and C are complementary to one another, while forming a base pair with three hydrogen bonds. Consequently, one may characterize AT base pairs as weak bases and GC base pairs as strong bases. In addition, the frequency of A(G) on a single strand is approximately equal to the frequency of T(C) on the same strand, a phenomenon that has been termed "strand symmetry" or "Chargaff's second parity". Therefore, DNA sequences can be transformed into sequences of weak W (A or T) and strong S (G or C) bases (Li and Holste, 2004). The SW mapping rule is particularly appropriate to analyze genome-wide correlations; this rule corresponds to the most fundamental partitioning of the four bases into their natural pairs in the double helix (G+C, A+T). The composition of base pairs, or GC level, is thus a strand-independent property of a DNA molecule and is related to important physico-chemical properties of the chain (Bernaola-Galvan *et al.*, 2002). The C+G content (isochore) studies have been done earlier (Bernardi *et al.*, 1985; Ikemura, 1985; Ikemura and Aota, 1988; Bernardi, 1989). The full story of how DNA really functions is not merely what is written on the sequence of base pairs. The DNA functions involve information transmission over many length scales ranging from a few to several hundred nanometers (Ball, 2003).

One of the major goals in DNA sequence analysis is to gain an understanding of the overall organization of the genome, in particular, to analyze the properties of the DNA string itself. Long-range correlations in DNA base sequence structure, which give rise to $1/f$ spectra have been identified (Fukushima *et al.*, 2002; Azad *et al.*, 2002). Such long-range correlations in space-time fluctuations is very common in nature and Li (2007) has given an extensive and informative bibliography of the observed $1/f$ noise or $1/f$ spectra, where $f$ is the frequency, in biological, physical, chemical and other dynamical systems. The long-range correlations in nucleotide sequence could in principle be explained by the coexistence of many different length scales. The advantage of spectral analysis is to reveal patterns hidden in a direct correlation function. The quality of the $1/f$ spectra differs greatly among sequences. Different DNA sequences do not exhibit the same power spectrum.

The concentration of genes is correlated with the C+G density. The spatial distribution of C+G density can be used to give an indication of the location of genes. The final goal is to eventually learn the "genome organization principles" (Li, 1997). The coding sequences of most verte-brate genes are split into segments (exons) which are separated by non-coding intervening sequences (introns). A very small minority of human genes lack noncoding introns and are very small genes (Strachan and Read, 1996).

Li (2002) reports that spectral analysis shows that there are GC con-tent fluctuations at different length scales in isochore (relatively homoge-neous) sequences. Fluctuations of all size scales coexist in a hierarchy of domains within domains (Li *et al.*, 2003). Li and Holste (2005) have recently identified universal $1/f$ spectra and diverse correlation structures in Guanine (G) and Cytosine (C) content of all human chromosomes.

In the following, it is shown that the frequency distribution of human chromosomes 1–24 DNA bases C+G concentration per 10 bp (non-over-lapping) follows the model prediction (Chapter 2) of self-organized criti-cality or quantum-like chaos implying long-range spatial correlations in the distribution of bases C+G along the DNA base sequence.

## 3.4 Data and Analysis

### 3.4.1 *Data*

The human chromosomes 1–24 DNA base sequence was obtained from the entrez Databases, Homo sapiens Genome (build 36 Version 1) at http://www.ncbi.nlm.nih.gov/entrez. The total number of contiguous data sets, each containing a minimum of 70,000 base pairs, chosen for the study are given in Fig. 3.3 for the chromosomes 1–24.

### 3.4.2 *Power spectral analyses: Variance and phase spectra*

The number of times base C and also base G, i.e., (C+G), occur in succes-sive blocks of 10 bases were determined in successive length sections of 70,000 base pairs giving a C+G frequency distribution series of 7,000 values for each data set. The power spectra of frequency distribution of C+G bases (per 10 bp) in the data sets were computed accurately by an elementary, but very powerful method of analysis developed by Jenkinson (1977) which provides a quasicontinuous form of the classical

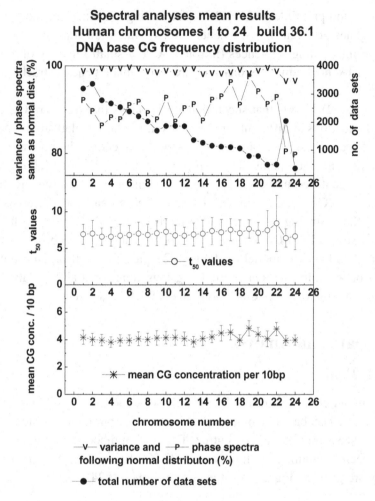

**Figure 3.3:** Average results of power spectral analyses. The error bars for one standard deviation are given for $t_{50}$ values and mean CG concentration per 10 bp.

periodogram allowing systematic allocation of the total variance and degrees of freedom of the data series to logarithmically spaced elements of the frequency range (0.5, 0). The cumulative percentage contribution to total variance was computed starting from the high frequency side of the spectrum. The power spectra were plotted as cumulative percentage contribution to total variance versus the normalized standard deviation $t$ equal to $(\log L/(\log T_{50})) - 1$ where $L$ is the period in years and $T_{50}$ is the period

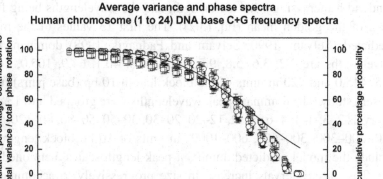

Average variance and phase spectra
Human chromosome (1 to 24) DNA base C+G frequency spectra

□ variance spectrum    ○ phase spectrum    —— normal distribution
The average variance and space spectra for each of the chromosomes 1 to 24
with error bars (one standard deviation) are plotted in the figure

**Figure 3.4:**   The average variance and phase spectra of frequency distribution of bases C+G in human chromosomes 1–24 for the data sets given in Fig. 3.3. The power spectra were computed as cumulative percentage contribution to total variance versus the normalized standard deviation $t$ equal to $(\log L/\log T_{50}) - 1$ where $L$ is the period in years and $T_{50}$ is the period up to which the cumulative percentage contribution to total variance is equal to 50. The corresponding phase spectra were computed as the cumulative percentage contribution to total rotation (Section 2.6).

up to which the cumulative percentage contribution to total variance is equal to 50 (Eq. (3.3)). The corresponding phase spectra were computed as the cumulative percentage contribution to total rotation (Section 3.2.6). The statistical chi-square test (Spiegel, 1961) was applied to determine the "goodness of fit" of variance and phase spectra with statistical normal distribution. Details of data sets and results of power spectral analyses are given in Fig. 3.3 as averages for each of the 24 chromosomes. The average variance and phase spectra for the data sets in each of the 24 chromosomes (Fig. 3.3) are given in Fig. 3.4.

### 3.4.3 *Power spectral analyses: Dominant periodicities*

The general systems theory predicts that the broadband power spectrum of fractal fluctuations will have embedded dominant wavebands, the bandwidth increasing with wavelength, and the wavelengths being functions of the golden mean (Eq. (3.2)). The first 13 values of the model predicted (Selvam, 1990; Selvam and Fadnavis, 1998) dominant peak wavelengths are 2.2, 3.6, 5.8, 9.5, 15.3, 24.8, 40.1, 64.9, 105.0, 167.0, 275, 445.0 and 720 in units of the block length 10 bp (base pairs) in the present study. The dominant peak wavelengths were grouped into 13 class intervals 2–3, 3–4, 4–6, 6–12, 12–20, 20–30, 30–50, 50–80, 80–120, 120–200, 200–300, 300–600, 600–1000 (in units of 10 bp block lengths) to include the model predicted dominant peak length scales mentioned earlier. The class intervals increase in size progressively to accommodate model predicted increase in bandwidth associated with increasing wavelength. Average class interval-wise percentage frequencies of occurrence of dominant wavelengths (normalized variance greater than 1) are shown in Fig. 3.5 along with the percentage contribution to total variance in each class interval corresponding to the normalized standard deviation $t$ (Eq. (3.3)) computed from the average $T_{50}$ (Fig. 3.3) for each of the 24 chromosomes. In this context, it may be mentioned that statistical normal probability density distribution represents the eddy variance (Eq. (3.3)). The observed frequency distribution of dominant eddies follow closely the computed percentage contribution to total variance.

### 3.4.3.1 *Peak wavelength versus bandwidth*

The model predicts that the apparently irregular fractal fluctuations contribute to the ordered growth of the quasiperiodic Penrose tiling pattern with an overall logarithmic spiral trajectory such that the successive radii lengths follow the Fibonacci mathematical series. Conventional power spectral analyses resolves such a spiral trajectory as an eddy continuum with embedded dominant wavebands, the bandwidth increasing with wavelength. The progressive increase in the radius of the spiral trajectory generates the eddy bandwidth proportional to the increment $d\theta$ in phase angle equal to $r/R$. The relative eddy circulation speed $W/w_*$ is directly proportional to the relative peak wavelength ratio $R/r$ since the eddy circulation speed $W = 2\pi R/T$, where $T$ is the eddy time period. The relationship between the peak wavelength and the bandwidth is obtained from Eq. (3.1), namely

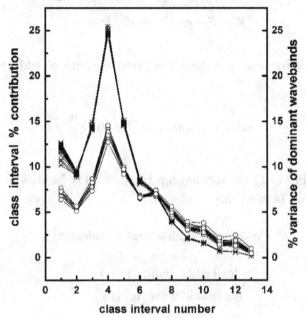

**Spectral analysis
DNA base C+G frequency distribution
Human chromosomes 1 to 24 build 36.1
Average distribution of dominant wavebands**

—∗— % no. of dominant wavebands
—○— % variance of dominant wavebands-
computed from mean $t_{50}$ for each chromosome

**Figure 3.5:** Dominant wavelengths in DNA bases C+G concentration distribution. Average class interval-wise percentage frequency distribution of dominant (normalized variance greater than 1) wavelengths is given by *line + star*. The corresponding computed percentage contribution to the total variance for each class interval is given by *line + open circle*. The observed frequency distribution of dominant eddies closely follow the model predicted computed percentage contribution to total variance.

$$W^2 = \frac{2}{\pi} \frac{r}{R} w_*^2$$

Considering eddy growth with overall logarithmic spiral trajectory

$$relative \; eddy \; bandwidth \propto d\theta \propto \frac{r}{R}$$

The eddy circulation speed is related to eddy radius as

$$W = \frac{2\pi R}{T}$$

$$W \infty R \infty peak \ \ wave \ \ length$$

The relative peak wavelength is given in terms of eddy circulation speed as

$$relative \ \ peak \ \ wavelength \infty \frac{W}{w_*}$$

From Eq. (3.1) the relationship between eddy bandwidth and peak wavelength is obtained as

$$eddy \ bandwidth = peak \ wavelength^2$$

$$\frac{\log(eddy \ bandwidth)}{\log(peak \ wave \ length)} = 2$$

A log–log plot of peak wavelength versus bandwidth will be a straight line with a slope (bandwidth/peak wavelength) equal to 2. A log–log plot of the average values of bandwidth versus peak wavelength shown in Fig. 3.6 exhibits average slopes approximately equal to 2.5.

## 3.5 Discussions

In summary, a majority of the data sets (Fig. 3.3) exhibit the model predicted quantum-like chaos for fractal fluctuations since the variance and phase spectra follow each other closely and also follow the universal inverse power law form of the statistical normal distribution signifying long-range correlations or coherence in the overall frequency distribution pattern of the DNA bases C+G in human chromosomes 1–24. Such non-local connections or "memory" in the spatial pattern is a natural

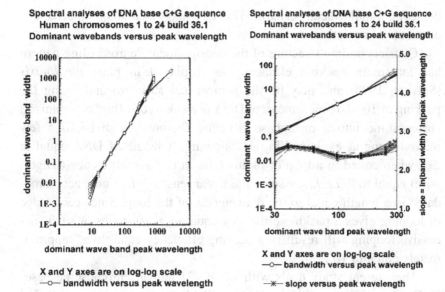

Spectral analyses of DNA base C+G sequence
Human chromosomes 1 to 24 build 36.1
Dominant wavebands versus peak wavelength

**Figure 3.6:** (left) The linear relationship between logarithms of dominant wave band-width versus corresponding peak wavelength. (right) shows the slope for a limited range of peak wavelength.

consequence of the model predicted Fibonacci spiral enclosing the space filling quasicrystalline structure of the quasiperiodic Penrose tiling pattern for fractal fluctuations of dynamical systems. Further, the broadband power spectra exhibit dominant wavelengths closely corresponding to the model predicted (Figs. 3.1 and 3.4; Eq. (3.2)) nested continuum of eddies. The apparently chaotic fluctuations of the frequency distribution of the DNA bases C+G per 10 bp in the human chromosomes 1–24 self-organize to form an ordered hierarchy of spirals or loops.

Analysis of auto-correlations of human chromosomes 1–22 and rice chromosomes 1–12 for seven binary mapping rules shows that the correlation patterns are different for different rules but almost identical for all of the chromosomes, despite their varying lengths (Podobnik *et al.*, 2007).

Mansilla *et al.* (2004) calculated the mutual information function for each of the 24 chromosomes in the human genome. The same correlation pattern is observed regardless the individual functional features of each chromosome. Moreover, correlations of different scale length are detected depicting a multifractal scenario. This fact suggests a unique mechanism of structural evolution.

Quasicrystalline structure of the quasiperiodic Penrose tiling pattern has maximum packing efficiency as displayed in plant *phyllotaxis* (Selvam, 1998) and may be the geometrical structure underlying the packing of $10^3$–$10^5$ micrometer of DNA in a eukaryotic (higher organism) chromosome into a metaphase structure (before cell division) a few microns long as explained in the following. A length of DNA equal to $2\pi L$ when coiled in a loop of radius $L$ has a packing efficiency (length-wise) equal to $2\pi L/2L = \pi$ since the linear length $2\pi L$ is now accommodated in a length equal to the diameter $2L$ of the loop. Since each stage of looping gives a packing efficiency equal to $\pi$, 10 stages of such successive looping will result in a packing efficiency equal to $\pi^{10}$ approximately equal to $10^5$.

The present study deals with all the 24 human chromosome bases C+G concentration per 10 bp in all available contiguous sequences. The window length 10 bp was chosen since the primary loop in the DNA molecule is equal to about 10 bp. The power spectral analysis gives the dominant wavelengths in terms of this basic unit, namely the window length of 10 bp. Increasing the window length (more than 10 bp) will result in decrease in resolution of shorter wavelengths. The aim of this preliminary study is to determine the spatial organization of the DNA bases C+G by applying concepts of a general systems theory first developed for atmospheric flows.

The important results of the present study are as follows: (1) the concentration per 10 bp of DNA bases C+G follow self-similar fractal fluctuations, namely an irregular series of successive increase followed by decrease on all size scales. (2) The power spectra of C+G concentration distribution follow the inverse power law form of the statistical normal distribution signifying quasicrystalline structure of the quasiperiodic Penrose tiling pattern for the spatial distribution of DNA bases C+G. (3) The quasiperiodic Penrose tiling pattern provides maximum packing efficiency for the DNA molecule inside the chromosome. (4) The observed inverse power law form for power spectra implies that the DNA bases are arranged in a fuzzy logic network with inherent long-range correlations.

# 3.6 Conclusions

Real world and model dynamical systems exhibit long-range space-time correlations, i.e., coherence, recently identified as self-organized criticality. Macroscale coherent functions in biological systems develop from self-organization of microscopic scale information flow and control such as in the neural networks of the human brain and in the His–Purkinje fibres of human heart, which govern vital physiological functions.

A recently developed cell dynamical system model for turbulent fluid flows predicts self-organized criticality as intrinsic to quantum-like mechanics governing flow dynamics. The model concepts are independent of exact details (physical, chemical, biological, etc.) of the dynamical system and are universally applicable. The model is based on the simple concept that space-time integration of microscopic domain fluctuations occur on self-similar fractal structures and give rise to the observed space-time coherent behaviour pattern with implicit long-term memory. Self-similar fractal structures to the spatial pattern for dynamical systems function as dynamic memory storage device with memory recall and update at all time scales.

The important conclusions of this study are as follows: (1) the frequency distribution of bases C+G per 10 bp in all the 24 human chromosomes DNA exhibit self-similar fractal fluctuations which follow the universal inverse power law form of the statistical normal distribution (Fig. 3.4), a signature of quantum-like chaos. (2) Quantum-like chaos indicates long-range spatial correlations or "memory" inherent to the self-organized fuzzy logic network of the quasiperiodic Penrose tiling pattern (Eq. (3.1) and Fig. 3.1). (3) Such non-local connections indicate that coding exons together with non-coding introns contribute to the effective functioning of the DNA molecule as a unified whole. Recent studies indicate that non-coding DNA introduce modifications in gene activity (Cohen, 2002; Makalowski, 2003). Studies now indicate that non-coding DNA may be responsible for the signals that were crucial for human evolution, directing the various components of our genome to work differently from the way they do in other organisms (Check, 2006). (4) The space filling quasiperiodic Penrose tiling pattern provides maximum packing efficiency for the DNA molecule inside the chromosome.

# Acknowledgement

The author is grateful to Dr. A. S. R. Murty for encouragement.

# References

Azad, R. K., Subba Rao, J., Li, W., and Ramaswamy, R. (2002). Simplifying the mosaic description of DNA sequences, *Phys. Rev. E*, 66, 031913 (1–6). http://www.nslij-genetics.org/wli/pub/pre02.pdf.

Bak, P., Tang, C., and Wiesenfeld, K. (1988). Self-organized criticality, *Phys. Rev. A*, 38, pp. 364–374.

Ball, P. (2003). Portrait of a molecule, *Nature*, 421, pp. 421–422.

Bernaola-Galvan, P., Carpena, P., Roman-Roldan, R. and Oliver, J. L. (2002). Study of statistical correlations in DNA sequences, *Gene*, 300(1–2), pp. 105–115. http://www.nslij-genetics.org/dnacorr/bernaola02.pdf.

Bernardi, G., Olofsson, B., Filipski, J., Zerial, M., Salinas, J., Cuny, G., Meunier-Rotival, M. and Rodier F. (1985). The mosaic genome of warm-blooded vertebrates, *Science*, 228, pp. 953–958.

Bernardi, G. (1989). The isochore organization of the human genome, *Annu. Rev. Genet.*, 23, pp. 637–661.

Berry, M. V. (1988). The geometric phase, *Sci. Amer.*, Dec., pp. 26–32.

Bertalanffy, L. Von (1968). *General Systems Theory: Foundations, Development, Applications* (George Braziller, New York).

Buchman, T. G. (2002). The community of the self, *Nature*, 420, pp. 246–251.

Check, E. (2006). It's the junk that makes us human, *Nature*, 444, pp. 130–131.

Cohen, P. (2002). New genetic spanner in the works, *New Scientist*, 16 March, p. 17.

Curl, R. F. and Smalley, R. E. (1991). Fullerenes, *Sci. Am.* (Indian Edition), 3, pp. 32–41.

Fukushima, A., Ikemura, T., Kinouchi, M., Oshima, T., Kudo, Y., Mori, H., and Kanaya, S. (2002). Periodicity in prokaryotic and eukaryotic genomes identified by power spectrum analysis, *Gene*, 300, pp. 203–211. http://www.nslij-genetics.org/dnacorr/fukushima02_gene.pdf.

Goldberger, A. L., Amaral, L. A. N., Hausdorff, J. M., Ivanov, P. Ch., Peng, C.-K., and Stanley, H. E. (2002). Fractal dynamics in physiology: Alterations with disease and aging, *PNAS*, 99, Suppl. 1, pp. 2466–2472. http://pnas.org/cgi/doi/10.1073/pnas.012579499.

Ikemura, T. (1985). Codon usage and tRNA content in unicellular and multicellular organisms, *Mol. Biol. Evol.*, 2, pp. 13–34.

Ikemura, T. and Aota, S. (1988). Global variation in G + C content along vertebrate genome DNA: Possible correlation with chromosome band structures, *J. Mol. Biol*, 203, pp. 1–13.

Janssen, T. (1988). Aperiodic crystals: A contradictio in terminals ? *Phys. Rep.*, 168(2) , pp. 55–113.

Jean, R. V. (1994). *Phyllotaxis: A Systemic Study in Plant Morphogenesis* (Cambridge University Press, New York).

Jenkinson, A. F. (1977). A Powerful Elementary Method of Spectral Analysis for use with Monthly, Seasonal or Annual Meteorological Time Series, Meteorological Office, London, Branch Memorandum No. 57, pp. 1–23.

Kepler, T. B., Kagan, M. L., and Epstein, I. R. (1991). Geometric phases in dissipative systems, *Chaos*, 1, pp. 455–461.

Kitano, H. (2002). Computational systems biology, *Nature*, 420, pp. 206–210.

Kitano, H. (2004). Biological robustness, *Nat. Rev. Genet.*, 5, pp. 826– 837.

Klir, G. J. (1992). Systems science: A guided tour, *J. Biol. Sys.*, 1, pp. 27–58.

Li, W. (1997). The study of correlation structure of DNA sequences: A critical review, *Comput. Chem.*, 21(4), pp. 257–272. http://www.nslij-genetics.org/wli/pub/cc97.pdf.

Li, W. and Holste, D. (2004). Spectral analysis of Guanine and Cytosine concentration of mouse genomic DNA, *Fluct. Noise Lett.*, 4(3), pp. L453–L464. http://www.nslij-genetics.org/wli/pub/fnl04.pdf.

Li, W. and Holste, D. (2005). Universal $1/f$ noise, crossovers of scaling exponents, and chromosome-specific patterns of guanine-cytosine content in DNA sequences of the human genome, *Phys. Rev. E*, 71, pp. 041910-1–9. http://www.nslij-genetics.org/wli/pub/pre05.pdf.

Li, W. (2007). A bibliography on $1/f$ noise. http://www.nslij-genetics.org/wli/1fnoise.

Maddox, J. (1988). License to slang Copenhagen? *Nature*, 332, p. 581.

Makalowski, W. (2003). Not junk after all, *Science*, 300, pp. 1246–1247.

Mansilla, R., Del Castillo, N., Govezensky, T., Miramontes, P., José, M., and Cocho, G. (2004). Long-range correlation in the whole human genome, arXiv.org/q-bio/0402043.

Milotti, E. (2002). *1/f* noise: A pedagogical review. http://arxiv.org/abs/physics/0204033.

Peacocke, A. R. (1989). *The Physical Chemistry of Biological Organization* (Clarendon Press, Oxford).

Podobnik, B., Shao, J., Dokholyan, N. V., Zlatic, V., H. Stanley H. E., and Grosse, I. (2007). Similarity and dissimilarity in correlations of genomic DNA, *Physica A*, 373, pp. 497–502.

Prigogine, I. and Stengers, I. (1988). *Order Out of Chaos*, 3rd Ed. (Fontana Paperbacks, London).

Ruhla, C. (1992). *The Physics of Chance* (Oxford University Press, Oxford), p. 217.

Selvam, A. M. (1990). Deterministic chaos, fractals and quantumlike mechanics in atmospheric flows, *Can. J. Phys.*, 68, pp. 831–841. http://xxx.lanl.gov/html/physics/0010046.

Selvam, A. M. and Fadnavis, S. (1998). Signatures of a universal spectrum for atmospheric interannual variability in some disparate climatic regimes, *Meteorol. Atmos. Phys.*, 66, pp. 87–112. http://xxx.lanl.gov/abs/chao-dyn/9805028.

Selvam, A. M. (2001a). Quantumlike chaos in prime number distribution and in turbulent fluid flows, *APEIRON*, 8(3), pp. 29–64. http://redshift.vif.com/JournalFiles/V08NO3PDF/V08N3SEL.PDF   http://xxx.lanl.gov/html/physics/0005067.

Selvam, A. M. (2001b). Signatures of quantumlike chaos in spacing intervals of non-trivial Riemann Zeta zeros and in turbulent fluid flows, *APEIRON*, 8(4), pp. 10–40. http://xxx.lanl.gov/html/physics/0102028 http://redshift.vif.com/JournalFiles/V08NO4PDF/V08N4SEL.PDF.

Selvam, A. M. (2002). Quantumlike chaos in the frequency distributions of the bases A, C, G, T in Drosophila DNA, *APEIRON*, 9(4), pp. 103–148. http://redshift.vif.com/JournalFiles/V09NO4PDF/V09N4sel.pdf.

Selvam, A. M. (2003). Signatures of quantum-like chaos in Dow Jones Index and turbulent fluid flows, *APEIRON*, 10, pp. 1–28. http://arxiv.org/html/physics/0201006,   http://redshift.vif.com/JournalFiles/V10NO4PDF/V10N4SEL.PDF.

Selvam, A. M. (2004). Quantumlike chaos in the frequency distributions of bases A, C, G, T in Human chromosome1 DNA, *APEIRON*, 11(3), pp. 134–146. http://redshift.vif.com/JournalFiles/V11NO3PDF/V11N3SEL.PDF http://arxiv.org/html/physics/0211066.

Spiegel, M. R. (1961). *Statistics* (McGraw-Hill, New York), p. 359.

Standish, T. G. (2002). Rushing to judgment: functionality in noncoding or "junk" DNA, *ORIGINS*, 53, pp. 7–30

Strachan, T. and Read, A. P. (1996). *Human Molecular Genetics* (Bios Scientific Publishers, Oxfordshire), p. 597.

Townsend, A. A. (1956). *The Structure of Turbulent Shear Flow*, 2nd Ed. (Cambridge University Press, Cambridge).

Vitiello, G. (1992). Coherence and electromagnetic fields in living matter, *Nanobiology*, 1, pp. 221–228.

Wang F., Weber P., Yamasaki K., Havlin S., and Stanley, H. E. (2006). Statistical regularities in the return intervals of volatility, *Eur. Phys. J.*, B., Published Online DOI: 10.1140/epjb/e2006-00356-9 (4 October 2006).

West, B. J. (2004). Comments on the renormalization group, scaling and measures of complexity, *Chaos Solit. Fractals*, 20, pp. 33–44.

Wooley, J. C. and Lin, H. S. (2005) Illustrative Problem domains at the interface of computing and biology. In *Catalyzing Inquiry at the Interface of Computing and Biology*, eds. Wooley, J. C. and Lin, H. S., p. 329 (Computer Science and Telecommunications Board, The National Academies Press, Washington, D.C.).

# Chapter 4

# Quantum-like Chaos in the Frequency Distributions of Bases A, C, G, T in Human Chromosome 1 DNA*

## 4.1 Introduction

DNA topology is of fundamental importance for a wide range of biological processes (Bates and Maxwell, 1993). Since the topological state of genomic DNA is of importance for its replication, recombination and transcription, there is an immediate interest to obtain information about the supercoiled state from sequence periodicities (Herzel *et al.*, 1998, 1999). Identification of dominant periodicities in DNA sequence will help understand the important role of coherent structures in genome sequence organization (Chechetkin and Turygin, 1995; Widom, 1996). Li (2002) has discussed meaningful applications of spectral analyses in DNA sequence studies. Recent studies indicate that the DNA sequence of letters A, C, G and T exhibit the inverse power law form $1/f^\alpha$ frequency spectrum where $f$ is the frequency and $\alpha$ the exponent. It is possible, therefore, that the sequences have long-range order (Li, 1992; Selvam, 2002; Li, Marr and Kaneko, 1995; Audit *et al.*, 2001; Stanley *et al.*, 2000; Stanley, 2000; Audit *et al.*, 2002; Som *et al.*, 2001). Inverse power-law

---

* http://arxiv.org/html/physics/0211066 (2004). *APEIRON*, 11(3), pp. 134–146.

form for power spectra of fractal space-time fluctuations is generic to dynamical systems in nature and is identified as self-organized criticality (Bak *et al.*, 1987). A general systems theory developed by the author (Mary Selvam, 1990; Selvam and Fadnavis, 1998, 1999) predicts the following: (a) long-range correlations intrinsic to self-organized criticality are a signature of quantum-like chaos, (b) the power spectra follow the universal inverse power law form of the statistical normal distribution, (c) fractal fluctuations self-organize to form the quasicrystalline structure of the space filling quasiperiodic Penrose tiling pattern. Here it is shown that power spectra of frequency distributions of bases A, C, G, T in human chromosome 1 DNA exhibit self-organized criticality. DNA is a quasicrystal possessing maximum packing efficiency (Stewart, 1995) in a hierarchy of spirals or loops.

## 4.2  Model Concepts

Power spectra of fractal space-time fluctuations of dynamical systems such as fluid flows, stock market price fluctuations, heart beat patterns, etc., exhibit inverse power-law form identified as self-organized criticality (Bak *et al.*, 1987) and represent a self-similar eddy continuum. A general systems theory (Mary Selvam, 1990; Selvam and Fadnavis, 1998, 1999) developed by the author shows that such an eddy continuum can be visualized as a hierarchy of successively larger-scale eddies enclosing smaller-scale eddies. Since the large eddy is the integrated mean of the enclosed smaller eddies, the eddy energy (variance) spectrum follows the statistical normal distribution according to the Central Limit Theorem (Ruhla, 1992). Hence, the additive amplitudes of eddies, when squared, represent the probabilities, which is also an observed feature of the subatomic dynamics of quantum systems such as the electron or photon (Maddox, 1988, 1993; Rae, 1988). The long-range correlations intrinsic to self-organized criticality in dynamical systems are signatures of quantum-like chaos associated with the following characteristics: (a) The fractal fluctuations result from an overall logarithmic spiral trajectory with the quasiperiodic Penrose tiling pattern (Mary Selvam, 1990; Selvam and Fadnavis, 1998, 1999) for the internal structure. (b) Conventional continuous periodogram power spectral analyses of such spiral trajectories will reveal a continuum of wavelengths with progressive increase in phase. (c) The broadband power spectrum will have embedded dominant wavebands, the bandwidth increasing with wavelength, and the wavelengths being functions of the golden mean. The first 13 values of the model predicted (Mary Selvam, 1990; Selvam and Fadnavis, 1998, 1999) dominant peak wavelengths are 2.2, 3.6, 5.8, 9.5, 15.3, 24.8, 40.1, 64.9, 105.0, 167.0, 275, 445.0 and 720 in units of the block length 10 bp (base pairs). Wavelengths (or periodicities) close to the model predicted values have been reported in weather and climate variability (Selvam and Fadnavis, 1998), prime number distribution (Selvam, 2001a), Riemann zeta zeros (non-trivial) distribution (Selvam, 2001b), stock market economics (Sornette *et al.*, 1995). (d) The conventional power spectrum plotted as the variance versus the frequency in log–log scale will now represent the eddy probability density on logarithmic scale versus the standard deviation of the eddy fluctuations on linear scale since the logarithm of the eddy wavelength represents the standard deviation, i.e., the root mean square (r.m.s) value of the eddy fluctuations. The r.m.s. value of the eddy fluctuations can be represented in terms of statistical normal distribution as follows. A normalized standard deviation $t = 0$ corresponds to cumulative percentage probability density equal to 50 for the mean value of the distribution. For the overall logarithmic spiral circulation the logarithm of the wavelength represents the r.m.s. value of eddy fluctuations and the normalized standard deviation $t$ is defined for the eddy energy as

$$t = \frac{\log_e L}{\log_e T_{50}} - 1 \qquad (4.1)$$

The parameter $L$ in Eq. (4.1) is the wavelength and $T_{50}$ is the wavelength up to which the cumulative percentage contribution to total variance is equal to 50 and $t = 0$. The variable $\log_e T_{50}$ also represents the mean value for the r.m.s. eddy fluctuations and is consistent with the concept of the mean level represented by r.m.s. eddy fluctuations. Spectra of time series of fluctuations of dynamical systems, for example, meteorological parameters, when plotted as cumulative percentage contribution to total variance versus $t$ follow the model predicted universal spectrum (Selvam and Fadnavis, 1998).

## 4.3  Data and Analyses

The human chromosome 1 DNA base sequence was obtained from the entrez Databases, Homo sapiens Genome (build 30) at http://www.ncbi. nlm.nih.gov/entrez. The first 10 contiguous data sets consisting of a total number of 9,931,745 bases (Table 4.1) were scanned to give a total number of 280 unbroken data sets of length 35,000 bases each for the study. The number of times that each of the four bases A, C, G, T occur in successive blocks of 10 bases were determined giving four groups of 3,500 frequency sequence values for each data set.

The power spectra of frequency distribution of bases were computed accurately by an elementary, but very powerful method of analysis developed by Jenkinson (1977) which provides a quasicontinuous form of the classical periodogram allowing systematic allocation of the total variance and degrees of freedom of the data series to logarithmically spaced elements of the frequency range (0.5, 0). The cumulative percentage

Table 4.1:

| S. no | Accession number | Base pairs | |
|-------|------------------|------------|---------|
|       |                  | from       | to      |
| 1  | NT_004350.12 | 1 | 2112206 |
| 2  | NT_004321.12 | 1 | 1131467 |
| 3  | NT_004547.12 | 1 | 92989   |
| 4  | NT_034374.1  | 1 | 175802  |
| 5  | NT_022058.11 | 1 | 156602  |
| 6  | NT_021917.12 | 1 | 303717  |
| 7  | NT_028054.9  | 1 | 3621417 |
| 8  | NT_032967.2  | 1 | 80421   |
| 9  | NT_021937.12 | 1 | 2077871 |
| 10 | NT_034375.1  | 1 | 179253  |

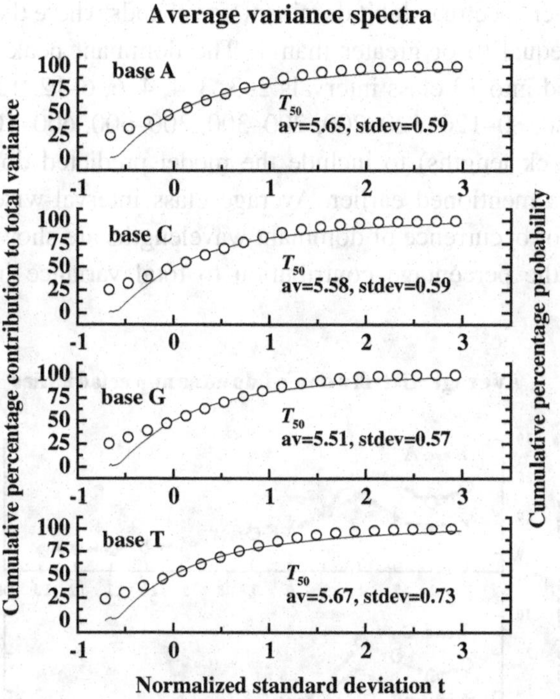

**Average variance spectra**

**Figure 4.1:** Average variance spectra for the four bases in human chromosome 1 DNA. Continuous lines are for the variance spectra and open circles give the statistical normal distribution. The mean and standard deviation of the wavelengths $T_{50}$ up to which the cumulative percentage contribution to total variance is equal to 50 are also given in the figure.

contribution to total variance was computed starting from the high frequency side of the spectrum. The power spectra were plotted as cumulative percentage contribution to total variance versus the normalized standard deviation $t$. The average variance spectra for the 280 data sets and the statistical normal distribution are shown in Fig. 4.1 for the four bases. The "goodness of fit" (statistical chi-square test) between the variance spectra and statistical normal distribution is significant at less than or equal to 5% level for 98.6%, 99.3%, 98.9% and 97.9% of the 280 data sets, respectively, for the four bases A, C, G and T. The average and standard deviation of the wavelength $T_{50}$ up to which the cumulative percentage contribution to total variance is equal to 50 are also shown in Fig. 4.1.

The power spectra exhibit dominant wavebands where the normalized variance is equal to or greater than 1. The dominant peak wavelengths were grouped into 13 class intervals 2–3, 3–4, 4–6, 6–12, 12–20, 20–30, 30–50, 50–80, 80–120, 120–200, 200–300, 300–600, 600–1000 (in units of 10 bp block lengths) to include the model predicted dominant peak length scales mentioned earlier. Average class interval-wise percentage frequencies of occurrence of dominant wavelengths are shown in Fig. 4.2 along with the percentage contribution to total variance in each class

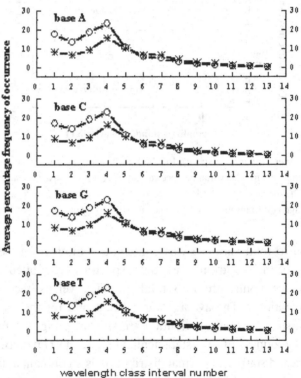

**Figure 4.2:**   Average wavelength class interval-wise percentage distribution of dominant (normalized variance greater than 1) wavelengths. Line + open circle give the average and dots denote one standard deviation on either side of the mean. The computed percentage contribution to the total variance for each class interval is given by line + star.

interval corresponding to the normalized standard deviation $t$ computed from the average $T_{50}$ (Fig. 4.1) for each of the four bases.

## 4.4 Results and Conclusions

The variance spectra for almost all the 280 data sets exhibit the universal inverse power-law form $1/f^{\alpha}$ of the statistical normal distribution (Fig. 4.1) where $f$ is the frequency and the spectral slope $\alpha$ decreases with increase in wavelength and approaches 1 for long wavelengths. The above result is also seen in Fig. 4.2 where the wavelength class interval-wise percentage frequency distribution of dominant wavelengths follow closely the corresponding computed variation of percentage contribution to the total variance as given by the statistical normal distribution. Inverse power-law form for power spectra implies long-range spatial correlations in the frequency distributions of the bases in DNA. Microscopic-scale quantum systems such as the electron or photon exhibit non-local connections or long-range correlations and are visualized to result from the superimposition of a continuum of eddies. Therefore, by analogy, the observed fractal fluctuations of the frequency distributions of the bases exhibit quantum-like chaos in the human chromosome 1 DNA. The eddy continuum acts as a robust unified whole fuzzy logic network with global response to local perturbations. Therefore, artificial modification of base sequence structure at any location may have significant noticeable effects on the function of the DNA molecule as a whole. Further, the presence of introns, which do not have meaningful code, may not be redundant, but may serve to organize the effective functioning of the coding exons in the DNA molecule as a complete unit

The results imply that the DNA base sequence self-organizes spontaneously to generate the robust geometry of logarithmic spiral with the quasiperiodic Penrose tiling pattern for the internal structure. The space filling geometric figure of the Penrose tiling pattern has intrinsic local five-fold symmetry (Devlin, 1997) and 10-fold symmetry. One of the three basic components of DNA, the deoxyribose is a five-carbon sugar and may represent the local five-fold symmetry of the quasicrystalline structure of the quasiperiodic Penrose tiling pattern of the DNA molecule as a whole. The DNA molecule shows 10-fold symmetry in the arrangement of 10 bases per turn of the double helix. The study of plant *phyllotaxis* in botany shows that quasicrystalline structure provides maximum packing efficiency for seeds, florets, leaves, etc. (Jean, 1994; Stewart, 1995; Mary Selvam, 1998). Quasicrystalline structure of the quasiperiodic *Penrose* tiling pattern may be the geometrical structure underlying the packing of $10^3$–$10^5$ micrometer of DNA in a eukaryotic (higher organism) chromosome into a metaphase structure a few microns long. The spatial geometry of the DNA is therefore organized into a hierarchy of helical structures. Such a concept may explain the observed loops of DNA in metaphase chromosome (Grosveld and Fraser, 1997). For example, the average class-interval wise percentage distribution of dominant periodicities show a peak in the wavelength interval 6–12 in units of 10 bp, i.e., 60–120 bp for all the four bases (Fig. 4.2). This predominant wavelength interval 60–120 bp may correspond to the coil length of each of the two DNA coils on the basic nucleosome unit of the chromatin fibre. Also, the value of $T_{50}$ ranges from 5 to 6 in units of 10 bp, i.e., from 50 to 60 bp (Fig. 4.1) indicating again the predominance of the fundamental coil length in the double coil of DNA in nucleosomes.

The general systems theory concepts applied to the observed frequency distribution of DNA base sequence in human chromosome 1 shows that the bases A, C, G, T self-organize to form the robust network architecture of the quasiperiodic Penrose tiling pattern. Buchman (2002) has given a detailed review of work done on systems theory as applied to biological systems and states that observed coupling mechanisms span multiple levels of resolution that demands bridging of molecular mechanisms and genetic data with physiological systems and functions.

## Acknowledgement

The author is grateful to Dr. A. S. R. Murty for encouragement.

## References

Audit, B. *et al.* (2001). Long-range correlations in genomic DNA: A signature of the nucleosomal structure, *Phys. Rev. Lett.*, 86(11), pp. 2471–2474. http://linkage.rockefeller.edu/wli/dna_corr/audit01.pdf.

Audit, B., Vaillant, C., Arneodo, A., d'Aubenton-Carafa, Y., and Thermes, C. (2002). Long-range correlations between DNA bending sites: Relation to the structure and dynamics of nucleosomes, *J. Mol. Biol.*, 316(4), pp. 903–918.

Bak, P., Tang, C., and Wiesenfeld, K. (1987). Self-organized criticality: An explanation of 1/*f* noise., *Phys. Rev. Lett.*, 59, pp. 381–384.

Bates, A. D. and Maxwell, A. (1993). *DNA Topology* (Oxford University Press, Oxford), p. 111.

Buchman, T. G. (2002). The community of the self, *Nature*, 420, pp. 246–251.

Chechetkin, V. R. and Yu. Turygin, A. (1995). Search of hidden periodicities in DNA sequences, *J. Theor. Biol.*, 175, 477–494. http://linkage.rockefeller.edu/wli/dna_corr/1995.html.

Devlin, K. (1997). *Mathematics: The Science of Patterns* (Scientific American Library, New York), p. 101.

Grosveld, F. and Fraser, P. (1997). Locus control of regions. In *Nuclear Organization, Chromatin Structure, and Gene Expression*, eds. Van Driel, R. and Otte, A. P., pp. 129–144 (Oxford University Press, Oxford).

Herzel, H., Weiss, O., and Trifonov, E. N. (1999). 10–11 bp periodicities in complete genomes reflect protein structure and DNA folding, *Bioinformatics*, 15(3), pp. 187–193. http://linkage.rockefeller.edu/wli/dna_corr/1999.html.

Herzel, H., Weiss, O., and Trifonov, E. N. (1998). Sequence periodicity in complete genomes of Archaea suggests positive supercoiling, *J. Biomol. Struct. Dynamics*, 16(2), pp. 341–345. http://linkage.rockefeller.edu/wli/dna_corr/1998.html.

Jean, R. V. (1994). *Phyllotaxis: A Systemic Study in Plant Morphogenesis* (Cambridge University Press, New York).

Jenkinson, A. F. (1977). *A Powerful Elementary Method of Spectral Analysis for Use with Monthly, Seasonal or Annual Meteorological Time Series*, Meteorological Office, London, Branch Memorandum No. 57, pp. 1–23.

Li, W. (2002). Are spectral analyses useful for DNA sequence analysis? *Proc. DNA in Chromatin, At the Frontiers of Biology, Biophysics, and Genomics*, 23–29 March, Arcachon, France. http://linkage.rockefeller.edu/wli/pub/arcachon02.pdf.

Li, W. (1992). Generating nontrivial long-range correlations and 1/*f* spectra by replication and mutation, *Int. J. Bifur. Chaos*, 2(1), pp. 137–154. http://linkage.rockefeller.edu/wli/dna_corr/l-ijbc92-l.html.

Li, W., Marr, T. G., and Kaneko, K. (1994). Understanding long-range correlations in DNA sequences, *Physica D*, 75(1–3), pp. 392–416; erratum: 82, p. 217 (1995). http://arxiv.org/chao-dyn/9403002.

Maddox, J. (1993). Can quantum theory be understood? *Nature*, 361, p. 493.

Maddox, J. (1988). Licence to slang Copenhagen? *Nature*, 332, p. 581.

Mary Selvam, A. (1990). Deterministic chaos, fractals and quantumlike mechanics in atmospheric flows, *Can. J. Phys.*, 68, pp. 831–841. http://xxx.lanl.gov/html/physics/0010046.

Mary Selvam, A. (1998). Quasicrystalline pattern formation in fluid substrates and phyllotaxis. In *Symmetry in Plants*, eds. Barabe, D. and Jean, R. V., Vol. 4. (World Scientific Series in Mathematical Biology and Medicine, Singapore), pp. 795–809. http://xxx.lanl.gov/abs/chao-dyn/9806001.

Rae, A. (1988). *Quantum-Physics: Illusion or Reality?* (Cambridge University Press, New York), p. 129.

Ruhla, C. (1992). *The Physics of Chance* (Oxford University Press, Oxford), p. 217.

Selvam, A. M. and Suvarna Fadnavis (1999). Superstrings, cantorian-fractal space-time and quantum-like chaos in atmospheric flows, *Chaos Solit Fractal.*, 10(8), pp. 1321–1334. http://xxx.lanl.gov/abs/chao-dyn/9806002.

Selvam, A. M. (2001). Quantum-like chaos in prime number distribution and in turbulent fluid flows, *APEIRON*, 8(3), pp. 29–64. http://redshift.vif.com/JournalFiles/V08NO3PDF/V08N3SEL.PDF http://xxx.lanl.gov/html/physics/0005067.

Selvam, A. M. (2002). Quantumlike chaos in the frequency distributions of the bases A, C, G, T in Drosophila DNA, *APEIRON*, 9(4), pp. 103–148. http://redshift.vif.com/JournalFiles/V09NO4PDF/V09N4sel.pdf.

Selvam, A. M. (2001). Signatures of quantum-like chaos in spacing intervals of non-trivial Riemann zeta zeros and in turbulent fluid flows, *APEIRON*, 8(4), pp. 10–40. http://redshift.vif.com/JournalFiles/V08NO4PDF/V08N4SEL.PDF http://xxx.lanl.gov/html/physics/0102028.

Selvam, A. M. and Fadnavis, S. (1998). Signatures of a universal spectrum for atmospheric interannual variability in some disparate climatic regimes, *Meteorol. Atmos. Phys.*, 66, pp. 87–112. http://xxx.lanl.gov/abs/chao-dyn/9805028.

Som, A., Chattopadhyay, Chakrabarti, J., and Bandyopadhyay, D. (2001). Codon distributions in DNA, *Phys. Rev. E*, 63, pp. 1–8. http://linkage.rockefeller.edu/wli/dna_corr/som01.pdf.

Sornette, D., Johansen, A., and Bouchaud, J.-P. (1995). Stock market crashes, precursors and replicas, *J. Phys.*, 6, 500018. http://xxx.lanl.gov/pdf/cond-mat/9510036.

Stanley, H. E., Amaral, L. A. N., Gopikrishnan, P., and Plerou, V. (2000). Scale invariance and universality of economic fluctuations, *Physica A*, 283, pp. 31–41.

Stanley, H. E. (2000). Exotic statistical physics: Applications to biology, medicine, and economics, *Physica A*, 285, pp. 1–17. http://www.elsevier.com/gej-ng/10/36/21/81/30/28/article.pdf.

Stewart, I. (1995). Daisy, daisy, give your answer do, *Sci. Amer.*, 272, pp. 76–79.

Widom, J. (1996). Short-range order in two eukaryotic genomes: Relation to chromosome structure, *J. Mol. Biol.*, 259, pp. 579–588. http://linkage.rockefeller.edu/wli/dna_corr/widom96.pdf.

# Chapter 5

# Universal Spectrum for DNA Base C+G Concentration Variability in Human Chromosome Y*

## 5.1 Introduction

Long-range space-time correlations, manifested as the self-similar fractal geometry to the spatial pattern, concomitant with inverse power law form for power spectra of space-time fluctuations are generic to spatially extended dynamical systems in nature and are identified as signatures of self-organized criticality. A representative example is the self-similar fractal geometry of His–Purkinje system whose electrical impulses govern the inter-beat interval of the heart. The spectrum of inter-beat intervals exhibits a broadband inverse power law form $f^\alpha$, where $f$ is the frequency and $\alpha$ the exponent. Self-organized criticality implies non-local connections in space and time, i.e., long-term memory of short-term spatial fluctuations in the extended dynamical system that acts as a unified whole communicating network.

*Nonlinear dynamics and chaos*, a multidisciplinary area of intensive research in recent years (since 1980s) has helped identify universal characteristics of spatial patterns (forms) and temporal fluctuations (functions)

* https://arxiv.org/ftp/physics/papers/0405/0405080.pdf. Proceedings of the WSEAS Multiconference: Mathematical Biology and Ecology (MABE'05), Udine, Italy, January 20–22, 2005.

of disparate dynamical systems in nature. Examples of dynamical systems, i.e., systems which change with time include biological (living) neural networks of the human brain which responds as a unified whole to a multitude of input signals and the non-biological (non-living) atmospheric flow structure which exhibits teleconnections, i.e., long-range space-time correlations. Spatially extended dynamical systems in nature exhibit self-similar fractal geometry to the spatial pattern. The sub-units of self-similar structures resemble the whole in shape. The name "fractal" coined by Mandelbrot (1977) indicates non-Euclidean or fractured (broken) Euclidean structures. Traditional Euclidean geometry discusses only three-, two- and one-dimensional objects, representative examples being sphere, rectangle and straight line, respectively. Objects in nature have irregular non-Euclidean shapes, now identified as fractals and the fractal dimension $D$ is given as $D = d\log M/d\log R$, where $M$ is the mass contained within a distance $R$ from a point within the extended object. A constant value for $D$ indicates uniform stretching on logarithmic scale. Objects in nature, in general exhibit multifractal structure, i.e., the fractal dimension $D$ varies with length scale $R$. The fractal structure of physiological systems has been identified (Goldberger *et al.*, 1990, 2002; West, 1990, 2004). The global atmospheric cloud cover pattern also exhibits self-similar fractal geometry (Lovejoy and Schertzer, 1986). Self-similarity implies long-range spatial correlations, i.e., the larger scale is a magnified version of the smaller scale with enhancement of fine structure. Self-similar fractal structures in nature support functions, which fluctuate on all scales of time. For example, the neural network of the human brain responds to a multitude of sensory inputs on all scales of time with long-term memory update and retrieval for appropriate global response to local input signals. Fractal architecture to the spatial pattern enables integration of a multitude of signals of all space-time scales so that the dynamical system responds as a unified whole to local stimuli, i.e., short-term fluctuations are carried as internal structure to long-term fluctuations. Fractal networks therefore function as dynamic memory storage devices, which integrate short-term fluctuations into long-period fluctuations. The irregular (nonlinear) variations of fluctuations in dynamical systems are therefore broadband because of the coexistence of fluctuations of all scales. Power spectral analysis (MacDonald, 1989) is conventionally used to resolve the periodicities (frequencies) and their amplitudes in time series data of fluctuations. The power spectrum is plotted on log–log scale as the intensity represented by variance (amplitude squared) versus the period (frequency) of the component periodicities. Dynamical systems in nature exhibit inverse power law form $1/f^{\alpha}$, where $f$ is the frequency (1/period) and $\alpha$ the exponent for the power spectra of space-time fluctuations indicating self-similar fluctuations on all space-time scales, i.e., long-range space-time correlations. The amplitudes of short term and long term are related by a scale factor alone, i.e., the space-time fluctuations exhibit scale invariance or long-range space-time correlations, which are independent of the exact details of dynamical mechanisms underlying the fluctuations at different scales. The universal characteristics of spatially extended dynamical systems, namely, the fractal structure to the space-time fluctuation pattern and inverse power law form for power spectra of space-time fluctuations are identified as signatures of self-organized criticality (Bak *et al.*, 1988). Self-organized criticality implies non-local connections in space and time in real world dynamical systems.

Surprisingly, such long-range space-time correlations had been earlier identified (Gleick, 1987) as sensitive dependence on initial conditions of finite precision computer realizations of nonlinear mathematical models of dynamical systems and named "deterministic chaos". Deterministic chaos is therefore a signature of self-organized criticality in computed model solutions.

It has not been possible to identify the exact mechanism underlying the observed universal long-range space-time correlations in natural dynamical systems and in computed solutions of model dynamical systems. The physical mechanisms responsible for self-organized criticality should be independent of the exact details (physical, chemical, physiological, biological, computational system, etc.) of the dynamical system so as to be universally applicable to all dynamical systems (real and model).

Atmospheric flows exhibit self-organized criticality as manifested in the fractal geometry to the global cloud cover pattern concomitant with inverse power law form for power spectra of temporal fluctuations documented and discussed in detail by Tessier *et al.* (1993). Standard models for atmospheric flow dynamics cannot explain the observed self-organized criticality in atmospheric flows satisfactorily. The author has developed a general systems theory for atmospheric flows (Selvam, 1990; Selvam and Fadnavis, 1998 and all references therein) that predicts the observed self-organized criticality as intrinsic to quantum-like chaos governing flow dynamics. The model concepts have also been applied to show that deterministic chaos in computed solutions of model dynamical systems is a direct consequence of round-off error in finite precision iterative computations (Selvam, 1993) incorporated in long-term numerical integration schemes used for numerical solutions.

In the following Section 5.2, the model for self-organized criticality in atmospheric flows is first summarized and model concepts are shown to be applicable to all real world and model dynamical systems. The concept of self-organized criticality and quantum-like chaos in biological and physiological systems in particular are discussed.

### 5.1.1 *General systems theory concepts*

In summary (Selvam, 1990; Selvam and Fadnavis, 1998), the model is based on Townsend's concept (Townsend, 1956) that large eddy structures form in turbulent flows as envelopes of enclosed turbulent eddies. Such a simple concept that space-time averaging of small-scale structures gives rise to large-scale space-time fluctuations leads to the following important model predictions.

### 5.1.2 *Quantum-like chaos in turbulent fluid flows*

Since the large eddy is but the integrated mean of enclosed turbulent eddies, the eddy energy (kinetic) distribution follows statistical normal distribution according to the *Central Limit Theorem* (Ruhla, 1992). Such a result, that the additive amplitudes of the eddies, when squared, represent probability distributions is found in the subatomic dynamics of quantum systems such as the electron or photon. Atmospheric flows, or, in general turbulent fluid flows follow quantum-like chaos.

### 5.1.3 *Dynamic memory (information) circulation network*

The root mean square (r.m.s.) circulation speeds $W$ and $w_*$ of large and turbulent eddies of respective radii $R$ and $r$ are related as

$$W^2 = \frac{2}{\pi}\frac{r}{R}w_*^2$$

$$(5.1)$$

Equation (5.1) is a statement of the law of conservation of energy for eddy growth in fluid flows and implies a two-way ordered energy flow between the larger and smaller scales. Microscopic scale perturbations are carried permanently as internal circulations of progressively larger eddies. Fluid flows therefore act as dynamic memory circulation networks with intrinsic long-term memory of short-term fluctuations. Such "memory of water" is reported by Davenas *et al.* (1988).

### 5.1.4 *Quasi-crystalline structure*

The flow structure consists of an overall logarithmic spiral trajectory with *Fibonacci* winding number and quasi-periodic *Penrose* tiling pattern for internal structure (Fig. 5.1). Primary perturbation $OR_O$ (Fig. 5.1) of time period $T$ generates return circulation $OR_1R_O$ which, in turn, generates successively larger circulations $OR_1R_2$, $OR_2R_3$, $OR_3R_4$, $OR_4R_5$, etc., such that the successive radii form the Fibonacci mathematical number series, i.e., $OR_1/OR_O = OR_2/OR_1 = \ldots\ldots = \tau$, where $\tau$ is the golden mean equal to $(1+\sqrt{5})/2 \approx 1.618$. The flow structure therefore consists of a nested continuum of vortices, i.e., vortices within vortices.

The quasiperiodic *Penrose* tiling pattern with five-fold symmetry has been identified as quasi-crystalline structure in condensed matter physics (Janssen, 1988). The self-organized large eddy growth dynamics, therefore, spontaneously generates an internal structure with the five-fold symmetry of the dodecahedron, which is referred to as the *icosahedral* symmetry, e.g., the *geodesic dome* devised by *Buckminster Fuller*. Incidentally, the pentagonal dodecahedron is, after the helix, nature's second favourite structure (Stevens, 1974). Recently the carbon macromolecule $C_{60}$, formed by condensation from a carbon vapour jet, was found to exhibit the icosahedral symmetry of the closed soccer ball and has been named *Buckminsterfullerene* or *footballene* (Curl and Smalley, 1991). Self-organized quasi-crystalline pattern formation therefore exists at the molecular level also and may result in condensation of specific biochemical structures in biological media. Logarithmic spiral formation with

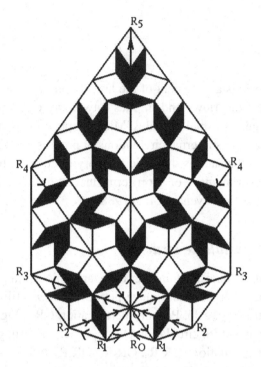

**Figure 5.1:**   The quasi-periodic Penrose tiling pattern which forms the internal structure at large eddy circulations.

*Fibonacci* winding number and five-fold symmetry possess maximum packing efficiency for component parts and are manifested strikingly in *Phyllotaxis* (Jean, 1992a, 1992b; 1994) and is common to nature (Stevens, 1974; Tarasov, 1986).

Model predicted spiral flow structure is seen vividly in the hurricane cloud cover pattern. Spiral waves are observed in many dynamical systems. Examples include Belousov–Zhabotinsky chemical reaction and also in the electrical activity of heart (Steinbock *et al.*, 1993).

### 5.1.5  *Dominant periodicities*

Dominant quasi-periodicities $P_n$ corresponding to the internal circulations (Fig. 5.1) $OR_0R_1$, $OR_1R_2$, $OR_2R_3$, ….. are given as

$$P_n = T(2 + \tau)\tau^n \qquad (5.2)$$

The dominant quasi-periodicities are equal to $2.2T$, $3.6T$, $5.8T$, $9.5T$, ......for values of $n = -1, 0, 1, 2,...$, respectively (Eq. (5.2)). Space-time integration of turbulent fluctuations results in robust broadband dominant periodicities which are functions of the primary perturbation time period $T$ alone and are independent of exact details (chemical, electrical, physical, etc.) of turbulent fluctuations. Also, such global scale oscillations in the unified network are not affected appreciably by failure of localized microscale circulation networks.

Wavelengths (or periodicities) close to the model predicted values have been reported in weather and climate variability (Selvam and Fadnavis, 1998), prime number distribution (Selvam, 2001a), Riemann zeta zeros (non-trivial) distribution (Selvam, 2001b), Drosophila DNA base sequence (Selvam, 2002), stock market economics (Selvam, 2003), Human chromosome 1 DNA base sequence (Selvam, 2004).

Similar unified communication networks may be involved in biological and physiological systems such as the brain and heart, which continue to perform overall functions satisfactorily in spite of localized physical damage. Structurally stable network configurations increase insensitivity to parameter changes, noise and minor mutations (Kitano, 2002).

Model predicted dominant quasi-periodicities (years) equal to 2.2, 3.6, 5.8, 9.5, (Eq. (5.2)) generated by the annual cycle ($T = 1$ year in Eq. (5.2)) of solar heating in atmospheric flows have been identified in global atmospheric weather patterns (Burroughs, 1992) as the quasi-biennial oscillation or QBO (2.2 years), the high frequency (3–4 years) and low frequency (5.8 years) components of the 3–7 years El Nino-Southern Oscillation (ENSO) cycle and decadic scale (>9 years) fluctuations. The ENSO cycle in particular is characterized by devastating regional changes in global climate pattern (Philander, 1990) and is now of public concern.

Persistent periodic energy pumping at fixed time intervals (period) $T$ in a fluid medium generates self-sustaining continuum of eddies and results in apparent nonlinear chaotic fluctuations in the fluid medium. Such chaotic optical (laser) emissions are triggered in nonlinear optical medium using a laser energy pump (Harrison and Biswas, 1986).

Self-organized broadband structures may therefore be generated in electromagnetic fields also.

Macroscale coherent structures emerge by space-time integration of microscopic domain fluctuations in fluid flows. Such a concept of the autonomous growth of atmospheric eddy continuum with ordered energy flow between the scales is analogous to Prigogine's (Prigogine and Stengers, 1988) concept of the spontaneous emergence of order and organization out of apparent disorder and chaos through a process of self-organization.

### 5.1.5.1 *Emergence of order and coherence in biology*

The problem of emergence of macroscopic variables out of microscopic dynamics is of crucial relevance in biology (Vitiello, 1992). In atmospheric flows turbulent fluctuations self-organize to form large eddies which give rise to cloud formations in updraft regions where moisture condenses. Similarly, in biological systems collective microscopic scale behaviour, e.g., self-organization of local information flow in neural networks may initiate global response in the human brain. Biological systems rely on a combination of network and the specific elements involved (Kitano, 2002). The notion that membership in a network could confer stability emerged from Ludwig von Bertalanffy's description of general systems theory in the 1930s and Norbert Wieners description of cybernetics in the 1940s. General systems theory focused in part on the notion of flow, postulating the existence and significance of flow equilibria. In contrast to Cannon's concept that mechanisms should yield homeostasis, general systems theory invited biologists to consider an alternative model of homeodynamics in which nonlinear, non-equilibrium processes could provide stability, if not constancy (Buchman, 2002).

The cell dynamical system model for coherent pattern formation in turbulent flows summarized earlier (Chapter 2) may provide a general systems theory for biological complexity. General systems theory is a logical-mathematical field, the subject matter of which is the formulation and deduction of those principles which are valid for "systems" in general, whatever the nature of their component elements or the relations or "forces" between them (Bertalanffy, 1968; Peacocke, 1989; Klir, 1993).

More than 25 years ago Frohlich (1968, 1970, 1975, 1980) introduced the concept of cooperative vibrational modes between proteins in biological cells. Coherent oscillations in the range of $10^{10}$–$10^{12}$ Hz involving cell membranes, DNA and cellular proteins could be generated by interaction between vibrating electric dipoles contained in the proteins as a result of nonlinear properties of the system. Through long-range effects proper to Frohlich nonlinear electrodynamics a temporospatial link, is in fact, established between all molecules constituting the system. Single molecules may thus act in a synchronized fashion and can no longer be considered as individual. New unexpected features arise from such a dynamic system, reacting as a unified whole entity (Insinnia, 1992). Coherent Frohlich oscillations may be associated with the dynamical pattern formation of intracellular cytoskeletal architecture consisting of networks of filamentous protein polymers, which coordinate and integrate information flow in the biological cell (Dayhoff *et al.*, 1994; Hameroff *et al.*, 1984, 1986, 1989; Hotani *et al.*, 1992; Jibu *et al.*, 1994). Grundler and Kaiser (1992), Kaiser (1992), Tabony and Job (1992) have also discussed biological auto-organization and pattern formation in the context of such coherent oscillations.

## 5.2 Long-Range Spatiotemporal Correlations (Coherence)

The logarithmic spiral flow pattern enclosing the vortices $OR_0R_1$, $OR_1R_2$, ... may be visualized as a continuous smooth rotation of the phase angle $\theta$ ($R_0OR_1$, $R_0OR_2$, ... etc.) with increase in period. The phase angle $\theta$ for each stage of growth is equal to $r/R$ and is proportional to the variance $W^2$ (Eq. (5.1)), the variance representing the intensity of fluctuations.

The phase angle gives a measure of coherence or correlation in space-time fluctuations. The model predicted continuous smooth rotation of phase angle with increase in period length associated with logarithmic spiral flow structure is analogous to Berry's phase (Berry, 1988; Kepler *et al.*, 1991) in quantum systems.

## 5.3   Universal Spectrum of Fluctuations

Conventional power spectral analysis will resolve such a logarithmic spiral flow trajectory as a continuum of eddies (broadband spectrum) with a progressive increase in phase angle.

The power spectrum, plotted on log–log scale as variance versus frequency (period) will represent the probability density corresponding to normalized standard deviation $t$ equal to $(\log L/\log T_{50}) - 1$ where $L$ is the period in years and $T_{50}$ is the period up to which the cumulative percentage contribution to total variance is equal to 50. The aforementioned expression for normalized standard deviation $t$ follows from model prediction of logarithmic spiral flow structure and model concept of successive growth structures by space-time averaging. Fluctuations of all scales therefore self-organize to form the universal inverse power law form of the statistical normal distribution. Since the phase angle $\theta$ equal to $r/R$ represents the variance $W^2$ (Eq. (5.1)), the phase spectrum plotted similar to variance spectrum will also follow the statistical normal distribution.

Signatures of quantum-like chaos, namely universal inverse power law form for atmospheric eddy energy spectrum and also model predicted quasi-periodicities associated with quasi-crystalline Penrose tiling pattern for internal flow structure (Fig. 5.1) have been identified in meteorological parameters (Selvam and Fadnavis, 1998).

## 5.4   Quantum Mechanics for Subatomic Dynamics: Apparent Paradoxes

The following apparent paradoxes found in the subatomic dynamics of quantum systems (Maddox, 1988) are consistent in the context of atmospheric flows as explained in the following section.

### 5.4.1   *Wave-particle duality*

A quantum system behaves as a wave on some occasions and as a particle at other times. Wave-particle duality is consistent in the context of

**WAVE - PARTICLE DUALITY**

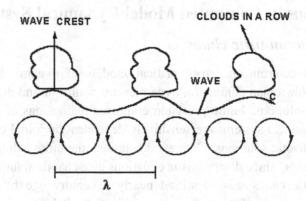

$\lambda$ = WAVE LENGTH    C = VELOCITY OF PROPAGATION

**Figure 5.2:**  Illustration of wave-particle duality as physically consistent for quantum-like mechanics in atmospheric flows. Particle-like clouds form in a row because of condensation of water vapour in updrafts and evaporation of condensed water in adjacent downdrafts associated with eddy circulations in atmospheric flows. Wave-particle duality in macroscale real world dynamical systems may be associated with bimodal (formation and dissipation) phenomenological form for manifestation of energy associated with bidirectional energy flow intrinsic to eddy (wave) circulations in the medium of propagation.

atmospheric waves, which generate particle-like clouds in a row because of formation of clouds in updrafts and dissipation of clouds in adjacent downdrafts characterizing wave motion (Fig. 5.2).

## 5.4.2 *Non-local connection*

The separated parts of a quantum system respond as a unified whole to local perturbations. Non-local connection is implicit to atmospheric flow structure quantified in Eq. (5.1) as ordered two-way energy flow between larger and smaller scales and seen as long-range space-time correlations, namely self-organized criticality. Atmospheric flows self-organize to form a unified network with the quasiperiodic *Penrose* tiling pattern for internal structure (Fig. 5.1), which provide long-range (non-local) space-time connections.

## 5.5 Self-Organized Criticality and Quantum-Like Chaos in Computed Model Dynamical Systems

### 5.5.1 *Deterministic chaos*

Traditional deterministic mathematical models of dynamical systems based on Newtonian continuum dynamics are nonlinear and do not have analytical solutions. Finite precision computer realizations of nonlinear model dynamical systems are sensitively dependent on initial conditions and give chaotic solutions. Computed solutions therefore exhibit "deterministic chaos" since deterministic equations give chaotic solutions. Such deterministic chaos was identified nearly a century ago by Poincare (1892) in his study of the three-body problem. Availability of computers with graphical display facilities in late 1950s facilitated numerical solutions and in 1963 Lorenz (1963) identified deterministic chaos in a simple model of atmospheric flows. Ruelle and Takens (1971) were the first to identify deterministic chaos as similar to turbulence in fluid flows. The computed trajectory traces the self-similar fractal pattern of the "strange attractor" so named because of its strange convoluted shape being the final destination of all possible trajectories. "Chaos Science" is now (since 1980s) an area of intensive research in all branches of science and other areas of human interest (Gleick, 1987). The physics of deterministic chaos is not yet identified. Deterministic chaos is a direct consequence of numerical solutions of error sensitive dynamical systems such as $X_{n+1}=F(X_n)$ where $X_{n+1}$, the value of the variable $X$ at the $(n+1)$th instant is a function $F$ of $X_n$. Error-feedback loop inherent to such iterative computations magnify exponentially with time the following errors inherent to numerical computations: (1) The continuous dynamical system is computed as a discrete dynamical system because of discretization of space and time in numerical computations with implicit assumption of sub-grid scale homogeneity. (2) Binary computer arithmetic precludes exact number representation at the data input stage itself. (3) Model approximations and assumptions. (4) Round-off error of finite precision computer arithmetic magnifies exponentially with time the above errors and gives chaotic solutions in iterative computations such as that used in long-term numerical integration schemes in numerical solutions.

Sensitive dependence on initial conditions of computed solutions indicates long-range space-time correlations and is a signature of self-organized criticality as explained earlier.

### 5.5.2  *Universal quantification for deterministic chaos in dynamical systems*

Selvam (1993) has shown that round-off error in finite precision computations is analogous to yardstick-length in length measurements. The computed domain at any stage of computation may be resolved as the product *WR* of the number of units of computation *W* of yardstick-length *R* and *wr* represents the initial uncertainty domain where *w* is the number of units of computation of precision *r* to begin with. Iterative computations may be visualized as spatial integration of enclosed higher precision uncertainty domain *wr* resulting in the larger uncertainty domain *WR*. The aforementioned concept is similar to the growth of large eddy structures from turbulent fluctuations. The concepts of cell dynamical system model for growth of large eddy structures in turbulent flows may therefore be applied for the growth of self-similar structures in the computed domain. The computed domain when resolved as a function of computational precision is shown (Selvam, 1993) to have an overall logarithmic spiral envelope with the quasiperiodic Penrose tiling pattern for the internal structure.

The computed dynamical system follows quantum-like mechanical laws with long-range space-time correlations manifested as the universal inverse power law form for power spectrum concomitant with fractal geometry to the spatial pattern. Deterministic chaos in computed dynamical systems is a manifestation of quantum-like mechanical laws governing round-off error flow dynamics with intrinsic non-local space-time connections, now identified as self-organized criticality.

### 5.5.3  *Universal algorithm for quasi-crystalline structure formation in real world and computed model dynamical systems*

Observed self-organized criticality in macroscale real world dynamical systems is a signature of quantum-like mechanics implemented in unified fractal structures which coordinate the cooperative existence of fluctuations of all space-time scales in the dynamical system.

Self-organized criticality in computed model dynamical systems, also is a result of quantum-like mechanical laws with ordered growth of round-off error structure similar to growth of large eddies from turbulent fluctuations in fluid flows. Such a concept may explain the surprising qualitative resemblance of patterns generated by computed dynamical systems to patterns in nature (Jurgen *et al.*, 1990; Stewart, 1992) with underlying universality quantified by the Fibonacci mathematical number series.

Generation of self-similar patterns by space-time integration of microscopic scale fluctuations underlie observed self-organized criticality in real world and computed dynamical systems.

The universal algorithm for self-organized criticality is identified as Eq. (5.1), which is the law of conservation of energy for space-time fluctuations. Equation (5.1) may be expressed in terms of universal Feigenbaum's constants $a$ and $d$ as (Selvam, 1993)

$$2a^2 = \pi d \tag{5.3}$$

Computed solutions of disparate dynamical systems exhibit period doublings which are quantified by two universal constants $a = -2.5029$ and $d = 4.6692$ named Feigenbaum's constants (Feigenbaum, 1980).

Eqution (5.3) states that the fractional volume intermittency of occurrence ($\pi d$) of fractal structures contributes to the total variance ($2a^2$) of the fluctuations (Selvam, 1993).

The physical mechanism underlying observed self-organized criticality in real world and computed model dynamical systems is quantified by Eq. (5.3), which is independent of the exact details (physical, chemical, biological, physiological, etc.), of dynamical systems and therefore applicable to all dynamical systems.

## 5.6 Applications of the General Systems Theory Concepts to Genomic DNA Base Sequence Structure

DNA sequences, the blueprint of all essential genetic information, are polymers consisting of two complementary strands of four types of bases:

adenine (A), cytosine (C), guanine (G) and thymine (T). Among the four bases, the presence of A on one strand is always paired with T on the opposite strand, forming a "base pair" with 2 hydrogen bonds. Similarly, G and C are complementary to one another, while forming a base pair with three hydrogen bonds. Consequently, one may characterize AT base-pairs as weak bases and GC base-pairs as strong bases. In addition, the frequency of A (G) on a single strand is approximately equal to the frequency of T (C) on the same strand, a phenomenon that has been termed "strand symmetry" or "Chargaff's second parity". Therefore, DNA sequences can be transformed into sequences of weak W (A or T) and strong S (G or C) bases (Li and Holste, 2004). The SW mapping rule is particularly appropriate to analyze genome-wide correlations; this rule corresponds to the most fundamental partitioning of the four bases into their natural pairs in the double helix (G+C, A+T). The composition of base pairs, or GC level, is thus a strand-independent property of a DNA molecule and is related to important physico-chemical properties of the chain (Bernaola-Galvan *et al.*, 2002). The full story of how DNA really functions is not merely what is written on the sequence of base-pairs; The DNA functions involve information transmission over many length scales ranging from a few to several hundred nanometers (Ball, 2003).

One of the major goals in DNA sequence analysis is to gain an understanding of the overall organization of the genome, in particular, to analyze the properties of the DNA string itself. Long-range correlations in DNA base sequence structure, which give rise to $1/f$ spectra have been identified (Azad *et al.*, 2002). Such long-range correlations in space-time fluctuations is very common in nature and Li (2004) has given an extensive and informative bibliography of the observed $1/f$ noise or $1/f$ spectra, where $f$ is the frequency, in biological, physical, chemical and other dynamical systems.

The long-range correlations in nucleotide sequence could in principle be explained by the coexistence of many different length scales. The advantage of spectral analysis is to reveal patterns hidden in a direct correlation function. The quality of the $1/f$ spectra differs greatly among sequences. Different DNA sequences do not exhibit the same power spectrum. The concentration of genes is correlated with the C+G density. The spatial distribution of C+G density can be used to give an indication of the location of genes. The final goal is to eventually learn the "genome organization principles" (Li, 1997). The coding sequences of most verte-brate genes are split into segments (exons) which are separated by non-coding intervening sequences (introns). A very small minority of human genes lack non-coding introns and are very small genes (Strachan and Read, 1996).

Li (2002) reports that spectral analysis shows that there are GC con-tent fluctuations at different length scales in isochore (relatively homoge-neous) sequences. Fluctuations of all size scales coexist in a hierarchy of domains within domains (Li *et al.*, 2003). Li and Holste (2004) have recently identified universal $1/f$ spectra and diverse correlation structures in guanine (G) and cytosine (C) content of all human chromosomes.

In the following section, it is shown that the frequency distribution of Human chromosome Y DNA bases C+G concentration per 10 bp (non-overlapping) follows the model prediction (Chapter 2) of self-organized criticality or quantum-like chaos implying long-range spatial correlations in the distribution of bases C+G along the DNA base sequence.

## 5.7 Data and Analysis

### 5.7.1 *Data*

The Human chromosome Y DNA base sequence was obtained from the entrez Databases, Homo sapiens Genome (build 34 Version 2) at http://www.ncbi.nlm.nih.gov/entrez. The 10 contiguous data sets containing a minimum of 70,000 base pairs chosen for the study are given in Table 5.1.

**Table 5.1:**   Data sets used for analyses.

| Set no. | Accession number | Base pairs | |
|---------|------------------|------------|----------|
|         |                  | From       | To       |
| 1 | NT_079581.1 | 1 | 86,563 |
| 2 | NT_079582.1 | 1 | 766,173 |
| 3 | NT_079583.1 | 1 | 623,707 |
| 4 | NT_079584.1 | 1 | 381,207 |
| 5 | NT_011896.8 | 1 | 6,323,261 |
| 6 | NT_011878.8 | 1 | 1,089,938 |

Table 5.1: (*Continued*)

| Set no. | Accession number | Base pairs | |
|---|---|---|---|
| | | From | To |
| 7 | NT_011875.10 | 1 | 9,938,763 |
| 8 | NT_011903.10 | 1 | 4,945,747 |
| 9 | NT_025975.2 | 1 | 98,295 |
| 10 | NT_079585.1 | 1 | 330,271 |

## 5.7.2 *Power spectral analyses: Variance and phase spectra*

The number of times base C and also base G, i.e., (C+G), occur in successive blocks of 10 bases were determined in successive length sections of 70,000 base pairs giving a C+G frequency distribution series of 7,000 values for each data set. The power spectra of frequency distribution of C+G bases in the data sets were computed accurately by an elementary, but very powerful method of analysis developed by Jenkinson (1977) which provides a quasi-continuous form of the classical periodogram allowing systematic allocation of the total variance and degrees of freedom of the data series to logarithmically spaced elements of the frequency range (0.5, 0). The cumulative percentage contribution to total variance was computed starting from the high frequency side of the spectrum. The power spectra were plotted as cumulative percentage contribution to total variance versus the normalized standard deviation $t$. The corresponding phase spectra were computed as the cumulative percentage contribution to total rotation (Section 2.6). The statistical chi-square test (Spiegel, 1961) was applied to determine the "goodness of fit" of variance and phase spectra with statistical normal distribution. Details of data sets and results of power spectral analyses are given in Table 5.2. The average variance and phase spectra for each of the 10 contiguous data sets (Table 5.1) are given in Fig. 5.3.

## 5.7.3 *Power spectral analyses: Dominant periodicities*

The general systems theory predicts the broadband power spectrum of fractal fluctuations will have embedded dominant wavebands, the

**Table 5.2:**    Results of power spectral analyses.

| Set no. | Base pairs used for analysis | | Number of data sets | Mean C+G concentration per 10bp | Mean $T_{50}$ | Spectra same as normal distribution (%) | |
|---|---|---|---|---|---|---|---|
| | From | To | | | | Variance | Phase |
| 1 | 1 | 70,000 | 1 | 5.47 | 6.75 | 100 | 100 |
| 2 | 1 | 700,000 | 10 | 4.79 | 10.20 | 100 | 90 |
| 3 | 1 | 560,000 | 8 | 4.87 | 8.71 | 100 | 100 |
| 4 | 1 | 350,000 | 5 | 4.61 | 7.36 | 100 | 100 |
| 5 | 1 | 6,300,000 | 90 | 3.82 | 6.22 | 96.7 | 78.9 |
| 6 | 1 | 1,050,000 | 15 | 4.13 | 7.98 | 66.7 | 73.3 |
| 7 | 1 | 9,870,000 | 141 | 3.68 | 6.34 | 98.6 | 83.7 |
| 8 | 1 | 4,900,000 | 70 | 3.95 | 6.58 | 95.7 | 68.6 |
| 9 | 1 | 70,000 | 1 | 3.89 | 3.48 | 100 | 0 |
| 10 | 1 | 280,000 | 4 | 3.86 | 5.94 | 100 | 50.0 |

bandwidth increasing with wavelength, and the wavelengths being functions of the golden mean (Section 2.4, Eq. (5.2)). The first 13 values of the model-predicted (Selvam, 1990; Selvam and Fadnavis, 1998) dominant peak wavelengths are 2.2, 3.6, 5.8, 9.5, 15.3, 24.8, 40.1, 64.9, 105.0, 167.0, 275, 445.0 and 720 in units of the block length 10 bp (base pairs) in the present study. The dominant peak wavelengths were grouped into 13 class intervals 2–3, 3–4, 4–6, 6–12, 12–20, 20–30, 30–50, 50–80, 80–120, 120–200, 200–300, 300–600, 600–1,000 (in units of 10 bp block lengths) to include the model predicted dominant peak length scales aforementioned. The class intervals increase in size progressively to accommodate model predicted increase in bandwidth associated with increasing wavelength. Average class interval-wise percentage frequencies of occurrence of dominant wavelengths are shown in Fig. 5.4 along with the percentage contribution to total variance in each class interval corresponding to the normalized standard deviation $t$ (Section 2.6) computed from the average $T_{50}$ (Table 5.2) for the 10 data sets.

## 5.8 Discussions

In summary, a majority of the data sets (Table 5.2 and Fig. 5.3) exhibit the model predicted quantum-like chaos for fractal fluctuations since the

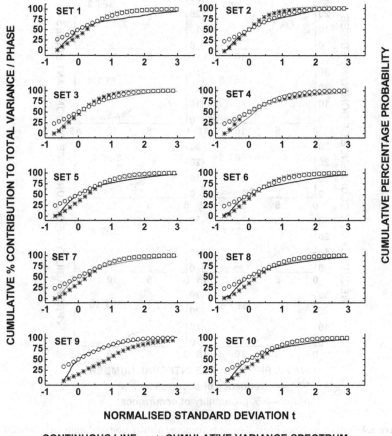

**Figure 5.3:** The average variance and phase spectra of frequency distribution of bases C+G in Human chromosome Y.

variance and phase spectra follow each other closely and also follow the universal inverse power law form of the statistical normal distribution signifying long-range correlations or coherence in the overall frequency distribution pattern of the bases C+G in Human chromosome Y DNA. Such non-local connections or "memory" in the spatial pattern is a natural consequence of the model predicted *Fibonacci* spiral enclosing the space filling quasi-crystalline structure of the quasiperiodic Penrose tiling

**WAVELENGTH CLASS INTERVAL-WISE DISTRIBUTION
OF DOMINANT EDDIES**

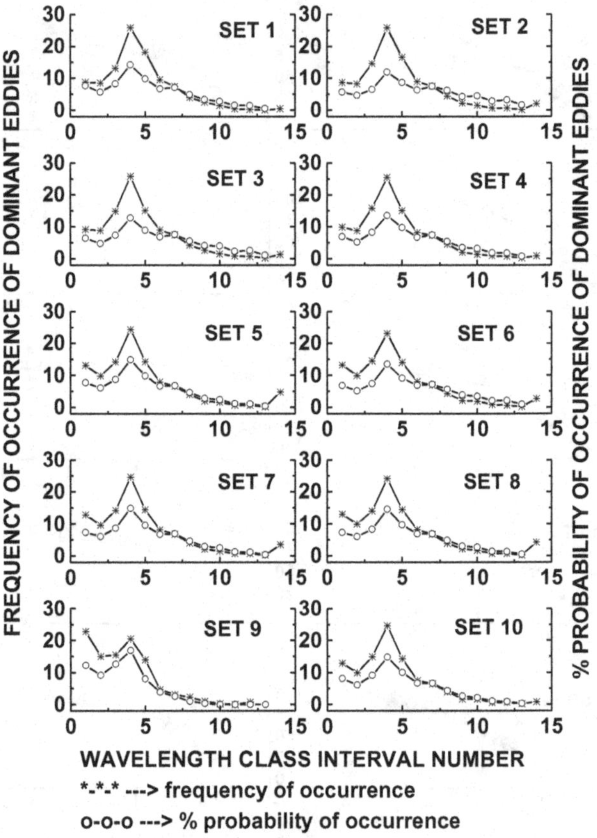

WAVELENGTH CLASS INTERVAL NUMBER
\*-\*-\* ---> frequency of occurrence
o-o-o ---> % probability of occurrence

**Figure 5.4:** Average class interval-wise percentage distribution of dominant (normalized variance greater than 1) wavelengths is given by *line + star*. The corresponding computed percentage contribution to the total variance for each class interval is given by *line + open circle*.

pattern for fractal fluctuations of dynamical systems. Further, the broad-band power spectra exhibit dominant wavelengths closely corresponding to the model predicted (Fig. 5.1 and Eq. (5.2)) nested continuum of eddies. The apparently chaotic fluctuations of the frequency distribution of the bases C+G per 10 bp in the Human chromosome Y DNA self-organize to form an ordered hierarchy of spirals or loops. Quasi-crystalline structure of the quasiperiodic Penrose tiling pattern has maximum packing efficiency as displayed in plant *phyllotaxis* (Selvam, 1998) and may be the geometrical structure underlying the packing of $10^3$–$10^5$ micrometer of DNA in a eukaryotic (higher organism) chromosome into a metaphase structure a few microns long as explained in the following. A length of DNA equal to $2\pi L$ when coiled in a loop of radius $L$ has a packing efficiency (lengthwise) equal to $2\pi L/2L = \pi$ since the linear length $2\pi L$ is now accommodated in a length equal to the diameter $2L$ of the loop. Since each stage of looping gives a packing efficiency equal to $\pi$, 10 stages of such successive looping will result in a packing efficiency equal to $\pi^{10}$ approximately equal to $10^5$.

## 5.9 Conclusions

Real world and model dynamical systems exhibit long-range space-time correlations, i.e., coherence, recently identified as self-organized criticality. Macroscale coherent functions in biological systems develop from self-organization of microscopic scale information flow and control such as in the neural networks of the human brain and in the His–Purkinje fibers of human heart, which govern vital physiological functions.

A recently developed cell dynamical system model for turbulent flows predicts self-organized criticality as intrinsic to quantum-like mechanics governing flow dynamics. The model concepts are independent of exact details (physical, chemical, biological, etc.) of the dynamical system and are universally applicable. The model is based on the simple concept that space-time integration of microscopic domain fluctuations occur on self-similar fractal structures and give rise to the observed space-time coherent behaviour pattern with implicit long-term memory. Self-similar fractal structures to the spatial pattern for dynamical systems function as dynamic memory storage device with memory recall and update at all-time scales.

Overall logarithmic spiral trajectory with quasiperiodic Penrose tiling pattern are intrinsic to dynamical systems. Such spiral architecture with *Fibonacci* winding number and five-fold symmetry are ubiquitous in plant kingdom (Jean, 1988, 1989, 1992a, 1992b, 1994) and are signatures of quantum-like chaos in macroscale dynamical systems. Simple laws underlie the exquisite varied patterns observed in nature.

The important conclusions of this study are as follows: (1) the frequency distribution of bases C+G per 10 bp in chromosome Y DNA exhibit self-similar fractal fluctuations which follow the universal inverse power law form of the statistical normal distribution (Fig. 5.3), a signature of quantum-like chaos. (2) Quantum-like chaos indicates long-range spatial correlations or "memory" inherent to the self-organized fuzzy logic network of the quasiperiodic Penrose tiling pattern (Eq. (5.1) and Fig. 5.1). (3) Such non-local connections indicate that coding exons together with non-coding introns contribute to the effective functioning of the DNA molecule as a unified whole. Recent studies indicate that mutations in introns introduce adverse genetic defects (Cohen, 2002). (4) The space filling quasiperiodic Penrose tiling pattern provides maximum packing efficiency for the DNA molecule inside the chromosome.

## Acknowledgement

The author is grateful to Dr. A. S. R. Murty for encouragement.

## References

Azad, R. K., Subba Rao, J., Li, W., and Ramaswamy, R. (2002). Simplifying the mosaic description of DNA sequences, *Phy. Rev. E*, 66, pp. 031913 (1–6). http://www.nslij-genetics.org/wli/pub/pre02.pdf.

Bak, P., Tang, C., and Wiesenfeld, K. (1988). Self-organized criticality, *Phys. Rev. A*, 38, pp. 364–374.

Ball, P. (2003). Portrait of a molecule, *Nature*, 421, pp. 421–422.

Bernaola-Galvan, P., Carpena, P., Roman-Roldan, R., and Oliver, J. L. (2002). Study of statistical correlations in DNA sequences, *Gene*, 300(1–2), pp. 105–115. http://www.nslij-genetics.org/dnacorr/bernaola02.pdf.

Berry, M. V. (1988). The geometric phase, *Sci. Amer. Dec.*, pp. 26–32.

Bertalanffy, L. Von (1968). *General Systems Theory: Foundations, Development, Applications* (George Braziller, New York).

Buchman, T. G. (2002). The community of the self, *Nature*, 420, pp. 246–251.

Burroughs, W. J. (1992). *Weather Cycles: Real or Imaginary?* (Cambridge University Press, Cambridge).

Cohen, P. (2002). New genetic spanner in the works, *New Scientist*, 16 March, p. 17.

Curl, R. F. and Smalley, R. E. (1991). Fullerenes, *Scientific American* (Indian Edition), 3, pp. 32–41.

Davenas, E., Beauvais, F., Amara, J., Oberbaum, M., Fortner, O., Belon, P., Sainte-Laudy, J., Robinzon, B., Miadonna, A., Tedeschi, A., Pomeranz, B., and Benvieste, J. (1988). Human basophil degranulation triggered by very dilute antiserum against IgE, *Nature*, 333, pp. 816–818.

Dayhoff, J., Hameroff, S., Lajoz-Beltra, R., and Swenberg, C. E. (1994). Cytoskeletal involvement in neuronal learning: A review, *Eur. Biophysics J.*, 23, pp. 79–93.

Feigenbaum, M. J. (1980). Universal behaviour in nonlinear systems, *Los Alamos Sci.*, 1, pp. 4–27.

Frohlich, H. (1968). Long-range coherence and energy storage in biological systems, *Int. J. Quantum Chem.*, 2, pp. 641–649.

Frohlich, H. (1970). Long-range coherence and the action of enzymes, *Nature*, 228, p. 1023.

Frohlich, H. (1975). The extraordinary dielectric properties of biological materials and the action of enzymes, *Proc. Natl. Acad. Sci. USA*, 72, pp. 4211–4215.

Frohlich, H. (1980). The biological effects of microwaves and related questions, *Adv. Electron. Electron. Phys.*, 53, pp. 85–152.

Gleick, J. (1987). *Chaos: Making a New Science* (Viking, New York).

Goldberger, A. L., Rigney, D. R., and West, B. J. (1990). Chaos and fractals in human physiology, *Sci. Am.*, Feb., pp. 42–49.

Goldberger, A. L., Amaral, L. A. N., Hausdorff, J. M., Ivanov, P. Ch., Peng, C.-K., and Stanley, H. E. (2002). Fractal dynamics in physiology: Alterations with disease and aging, *PNAS*, 99(1), pp. 2466–2472. http://pnas.org/cgi/doi/10.1073/pnas.012579499 .

Grundler, W. and Kaiser, F. (1992). Experimental evidence for coherent excitations correlated with cell growth, *Nanobiology*, 1, pp. 163–176.

Hameroff, S. R., Smith, S. A., and Watt, R. C. (1984). Nonlinear electrodynamics in cytoskeletal protein lattices. In *Nonlinear Electrodynamics in Biology and Medicine*, eds. Lawrence F. A. and Adey, W. R. (Plenum, New York).

Hameroff, S. R., Smith, S. A., and Watt, R. C. (1986). Automation model of dynamic organization in microtubules, *Ann. NY. Acad. Sci.*, 466, pp. 949–952.

Hameroff, S. R., Rasmussen, S., and Mansson, B. (1989). Molecular automata in MT: Basic computational logic of the living state? In *Artificial Life: Santa Fe*

*Instutute Studies in the Sciences of Complexity.* Vol. VI, ed. Langton, C. (Addison Wesley, Reading, MA).

Harrison, R. G. and Biswas, D. J. (1986). Chaos in light, *Nature*, 321, pp. 394–401.

Hotani, H., Lahoz-Beltra, R., Combs, B., Hameroff, S. R., and Rasmussen, S. (1992). Microtubule dynamics, liposomes and artificial cells: *In vitro* observation and cellular automata simulation of microtubule assembly and membrane morphogenesis, *Nanobiology*, 1, pp. 61–74.

Insinnia, E. M. (1992). Synchronicity and coherent excitations in microtubules, *Nanobiology*, 1(2), pp. 191–208.

Jean, R. V. (1992a). Nomothetical modelling of spiral symmetry in biology. In *Fivefold Symmetry*, ed. Hargittai I. (World Scientific, Singapore), pp. 505–528.

Jean, R. V. (1992b). On the origins of spiral symmetry in plants. In *Spiral Symmetry*, ed. Hargittai I. (World Scientific, Singapore), pp. 323–351.

Jean, R. V. (1994). *Phyllotaxis: A Systemic Study in Plant Morphogenesis* (Cambridge University Press, NY, USA).

Jenkinson, A. F. (1977). A Powerful Elementary Method of Spectral Analysis for use with Monthly, Seasonal or Annual Meteorological Time Series, Meteorological Office, London, Branch Memorandum No. 57, pp. 1–23.

Jibu, M., Hagan, S., Hameroff, S. R., Pribam, K. H., and Yasaue, K. (1994). Quantum optical coherence in cytoskeletal microtubules: Implications for brain function, *Biosystems*, 32(3), pp. 195–209.

Jurgen, H., Peitgen, H.-O., and Saupe, D. (1990). The language of fractals, *Sci. Amer.*, 263, pp. 40–49.

Kaiser, F. (1992). Biophysical models related to Frohlich excitations, *Nanobiology*, 1, pp. 163–176.

Kepler, T. B., Kagan, M. L., and Epstein, I. R. (1991). Geometric phases in dissipative systems, *Chaos*, 1, pp. 455–461.

Kitano, H. (2002). Computational systems biology, *Nature*, 420, pp. 206–210.

Klir, G. J. (1992). Systems science: A guided tour, *J. Biological Systems*, 1, pp. 27–58.

Li, W. (1997). The study of correlation structure of DNA sequences: A critical review, *Computers Chem.*, 21(4), pp. 257–272. http://www.nslij-genetics.org/wli/pub/cc97.pdf.

Li, W. (2002). Are isochore sequences homogeneous? *Gene*, 300(1–2), pp. 129–139. http://www.nslij-genetics.org/wli/pub/gene02.pdf.

Li, W., Bernaola-Galvan, P., Carpena, P., and Oliver, J. L. (2003). Isochores merit the prefix "iso", *Computational Biology and Chemistry*, 27, pp. 5–10. http://www.nslij-genetics.org/wli/pub/cbc03.pdf.

Li, W. (2004). A bibliography on $1/f$ noise. http://www.nslij-genetics.org/wli/1fnoise.

Li, W. and Holste, D. (2004). Spectral analysis of Guanine and Cytosine concentration of mouse genomic DNA, *Fluct. Noise Lett.*, 4(3), pp. L453–L464.

Li, W. and Holste, D. (2005). Universal $1/f$ spectra and diverse correlation structures in Guanine and Cytosine content of Human Chromosomes, *Phys. Rev. E* 71, pp. 041910 (9 pages).

Lorenz, E. N. (1963). Deterministic non-periodic flow, *J. Atmos. Sci.*, 20, pp. 130–141.

Lovejoy, S. and Schertzer, D. (1986). Scale invariance, symmetries, fractals and stochastic simulations of atmospheric phenomena, *Bull. Amer. Meteorol. Soc.*, 67, pp. 21–32.

Macdonald, G. F. (1989). Spectral analysis of time series generated by nonlinear processes, *Rev. Geophys.*, 27, pp. 449–469.

Maddox, J. (1988). License to slang Copenhagen? *Nature*, 332, p. 581.

Mandelbrot, B. B. (1977). *Fractals: Form, Chance and Dimension* (Freeman, San Francisco).

Peacocke, A. R. (1989). *The Physical Chemistry of Biological Organization* (Clarendon Press, Oxford, UK).

Philander, G. S. (1990). *El Nino, La Nina and the Southern Oscillation*, Int'l Geophysical Series 46 (Academic Press Inc., California).

Poincare, H. (1892). *Les Methodes Nouvelle de la Mecannique Celeste* (Gautheir-Villars, Paris).

Prigogine, I. and Stengers, I. (1988). *Order Out of Chaos*, 3rd Ed. (Fontana Paperbacks, London).

Ruelle, D. and Takens, F. (1971). On the nature of turbulence, *Commun. Math. Phys.*, 20, pp. 23, 167–192, 343–344.

Ruhla, C. (1992). *The Physics of Chance* (Oxford University Press, Oxford), pp. 217.

Selvam, A. M. (1990). Deterministic chaos, fractals and quantum-like mechanics in atmospheric flows, *Can. J. Phys.*, 68, pp. 831–841. http://xxx.lanl.gov/html/physics/0010046.

Selvam, A. M. (1993). Universal quantification for deterministic chaos in dynamical systems, *Appl. Math. Modelling*, 17, pp. 642–649. http://xxx.lanl.gov/html/physics/0008010.

Selvam, A. M. and Fadnavis, S. (1998). Signatures of a universal spectrum for atmospheric interannual variability in some disparate climatic regimes, *Meteorol. & Atmos. Phys.*, 66, pp. 87–112. http://xxx.lanl.gov/abs/chao-dyn/ 9805028.

Selvam, A. M. (2001a). Quantumlike chaos in prime number distribution and in turbulent fluid flows, *APEIRON*, 8(3), pp. 29–64. http://redshift.vif.com/ JournalFiles/V08NO3PDF/V08N3SEL.PDF http://xxx.lanl.gov/html/ physics/0005067.

Selvam, A. M. (2001b). Signatures of quantum-like chaos in spacing intervals of non-trivial Riemann Zeta zeros and in turbulent fluid flows, *APEIRON*, 8(4), pp. 10–40. http://xxx.lanl.gov/html/physics/0102028 http://redshift.vif.com/ JournalFiles/V08NO4PDF/V08N4SEL.PDF.

Selvam, A. M. (2002). Quantumlike chaos in the frequency distributions of the bases A, C, G, T in Drosophila DNA, *APEIRON*, 9(4), pp. 103–148. http:// redshift.vif.com/JournalFiles/V09NO4PDF/V09N4sel.pdf.

Selvam, A. M. (2003). Signatures of quantum-like chaos in Dow Jones Index and turbulent fluid flows. *APEIRON*, 10, pp. 1–28. http://arxiv.org/html/ physics/0201006. http://redshift.vif.com/JournalFiles/V10NO4PDF/ V10N4SEL.PDF.

Selvam, A. M. (2004). Quantumlike chaos in the frequency distributions of bases A, C, G, T in Human chromosome1 DNA, *APEIRON*, 11(3), pp. 134–146. http://arxiv.org/html/physics/0211066.

Spiegel, M. R. (1961). *Statistics* (McGraw-Hill, New York), p. 359.

Steinbock, O., Zykov, V., and Muller, S. C. (1993). Control of spiral-wave dynamics in active media by periodic modulation of excitability, *Nature*, 366, pp. 322–324.

Stevens, P. S. (1974). *Patterns in Nature* (Little, Brown and Co. Inc., Boston, USA).

Stewart, I. (1992). Where do nature's patterns come from? *New Scientist*, 135, p. 14.

Strachan, T. and Read, A. P. (1996). *Human Molecular Genetics* (βios Scientific Publishers), p. 597.

Tabony, J. and Job, D. (1992). Molecular dissipative structures in biological auto-organization and pattern formation, *Nanobiology*, 1, pp. 131–148.

Tarasov, L. (1986). *This Amazingly Symmetrical World* (Mir Publishers, Moscow).

Tessier, Y., Lovejoy, S., and Schertzer, D. (1993). Universal multifractals: Theory and observations for rain and clouds, *J. Appl. Meteorol.*, 32, pp. 223–250.

Townsend, A. A. (1956). *The Structure of Turbulent Shear Flow*, 2nd Ed. (Cambridge University Press).

Vitiello, G. (1992). Coherence and electromagnetic fields in living matter, *Nanobiology*, 1, pp. 221–228.

West, B. J. (1990). Fractal forms in physiology, *Int. J. Modern Physics B*, 4(10), pp. 1629–1669.

West, B. J. (2004). Comments on the renormalization group, scaling and measures of complexity, *Chaos Soliton. Fract.*, 20, pp. 33–44.

# Chapter 6

# Quantum-like Chaos in the Frequency Distributions of the Bases A, C, G, T in Drosophila DNA*

## 6.1 Introduction

### 6.1.1 *The DNA molecule and heredity*

Heredity in living organisms is determined by a long complex chemical molecule called deoxyribonucleic acid (DNA). The units of heredity, the genes, are parts of the DNA molecule situated along the length of the chromosomes inside the nucleus of the cell. A simplified picture of the molecule of DNA may be visualized to consist of two long backbones with projections sticking out from them at right angles rather like a ladder with its two upright sides and its rungs. The backbones are made up of two simple chemicals arranged alternately — *sugar — phosphate — sugar — phosphate* — all along the way. The projections are the four units or "letters" of the code; they are four chemical bases called *guanine, cytosine, adenine* and *thymine* — G, C, A, T. These four bases are arranged in a specific sequence which constitute the genetic code. The DNA molecule actually consists not of a single thread, but of two helical threads wound around each other — a double helix. The two DNA chains run in opposite directions and are coiled around each other with the bases facing one

---

* http://arxiv.org/html/physics/0210068 (2002). *APEIRON*, 9(4), pp. 103–148.

another in pairs. Only specific pairs of bases can be linked together, T always pairs with A, and G with C (Claire, 1964; Bates and Maxwell, 1993). The amount of A is the same as the amount of T, while the amount of G is the same as the amount of C. These are now known as Chargaff ratios (Gribbin, 1985; Alcamo, 2001).

What distinguishes one type of cell from another and one organism from another is the protein, which it contains. And it is DNA which dictates to the cell how many and what types of protein it shall make. Twenty different chemicals called amino acids in different sets of combinations form the proteins. The sequence of bases along each DNA molecule in the chromosome determines the sequence of amino acids along each of the proteins. It takes a sequence of three bases, the codon, to identify one amino acid. The order in which these bases recur within a particular gene in the helix corresponds to the information needed to build that gene's particular protein (Claire, 1964; Leone, 1992; Ball, 2000).

The genes of higher organisms are seldom "recorded" in the chromosomes intact, but are scattered in fragmentary fashion along a stretch of DNA, broken up by chunks of DNA which seem at first sight to carry no message at all. All the useless or "junk" DNA, the intervening sequences are known as *introns*. The pieces of DNA carrying genetic code are called *exons*. The codons, 64 in number are distributed over the coding parts of the DNA sequences. It is well known that the coding regions are translated into proteins. The non-coding parts are presumed important in regulatory and promotional activities. The biologically meaningful structures in noncoding regions are not known (Gribbin, 1985; Guharay *et al.*, 2000; Clark, 2001; Som *et al.*, 2001). Understanding genetic defects will make it easier to treat them (Watson, 1997).

Historically, Watson and Crick (1953) put together all the experimental data concerning DNA and decided that the only structure that fitted all the facts was the double helix and postulated that DNA is composed of two ribbon-like "backbones" composed of alternating deoxyribose and phosphate molecules. They surmised that nucleotides extend out from the backbone chains and that the 0.34 nm distance represents the space between successive nucleotides. The X-ray data showed a distance of 34 nm between turns, so they guessed that 10 nucleotides exist per turn. One strand of DNA would only encompass 1 nm width, so they postulated that

DNA is composed of two strands to conform to the 2 nm diameter observed in the X-ray diffraction photographs. Scientists now agree that DNA is arranged as a double helix of two intertwined chains, with complementary bases (A-T and G-C) opposing each other. Moreover, the strands run opposite to one another, that is, the strands display the reverse polarity. Given the base sequence of one chain of DNA, the base sequence of its partner chain is automatically determined by simply noting which bases are complimentary (*adenine-thymine* or *cytosine-guanine*). Furthermore, the structure provides a mechanism by which one chain can serve as a template (a model or pattern) for the synthesis of the other chain (Sambamurty, 1999; Alcamo, 2001). The genomic DNA in cells must be highly compacted in order to be contained in the required space. Each chromosome appears to contain a single giant molecule of DNA. At least three levels of condensation are required to package the $10^3$–$10^5$ micrometer of DNA in a eukaryotic (higher organism) chromosome into a metaphase structure a *few* microns long. The first level of condensation involves packaging DNA as a supercoil into nucleosomes. This produces 10 nm diameter interphase chromatin fibre. Second level of condensation involves an additional folding and/or supercoiling of the 10 nm nucleosome fibre to produce the 30 nm chromatin fibre. This third level of condensation appears to involve the segregation of segments of the giant DNA molecules present in eukaryotic chromosomes into independently supercoiled domains or loops. The mechanism by which this third level of condensation occurs is not known (Sambamurty, 1999).

## 6.1.2 *Long-range correlations in DNA base sequence*

DNA topology is of fundamental importance for a wide range of biological processes (Bates and Maxwell, 1993). One big question in DNA research is whether there is some meaning to the order of the base pairs in DNA. Human DNA has become a fascinating topic for physicists to study. One reason for this fascination is the fact that when living cells divide, the DNA is replicated exactly. This is interesting because approximately 95% of human DNA is called "junk" even by biologists who specialize in DNA. One practical task for physicists is simply to identify which sequences within the molecule are the coding sequences. Another scientific interest is to discover why the "junk" DNA is there in the first place. Almost everything in biology has a purpose that, in principle, is discoverable (Stanley, 2000). The study of statistical patterns in DNA sequences is important as it may improve our understanding of the organization and evolution of life on the genomic level. Recent studies indicate that the DNA sequence of letters A, C, G and T does have a $1/f^\alpha$ frequency spectrum where $f$ is the frequency and $\alpha$ the exponent. It is possible, therefore, that the sequences have long-range order and underlying grammar rules. The opinion on this issue remains divided (Som *et al.*, 2001 and all references therein). The findings of long-range correlations (LRC) in DNA sequences have attracted much attention, and attempts have been made to relate those findings to known biological features such as the presence of triplet periodicities in protein-coding DNA sequences, the evolution of DNA sequences, the length distribution of protein-coding regions, or the expansion of simple sequence repeats (Holste *et al.*, 2001).

A summary of recent results relating to LRC in DNA sequences is given in the following. Based on spectral analyses, Li *et al.* found (Li, 1992; Li and Kaneko, 1992; Li *et al.*, 1994) that the frequency spectrum of a DNA sequence containing mostly *introns* shows $1/f^\alpha$ behavior, which evidence the presence of LRC. The correlation properties of coding and non-coding DNA sequences were first studied by Peng *et al.* (1992) in their *fractal* landscape or DNA walk model. Peng *et al.* (1992) discovered that there exists LRC in non-coding DNA sequences while the coding sequences correspond to a regular random walk. By doing a more detailed analysis of the same data set, Chatzidimitriou-Dreismann and Larhammar (1993) concluded that both coding and non-coding sequences exhibit LRC. A subsequent work by Prabhu and Claverie (1992) also substantially corroborates these results. Buldyrev *et al.* (1995) showed the LRC appears mainly in non-coding DNA using all the DNA sequences available. Alternatively, Voss (1992, 1994), based on equal-symbol correlation, showed a power-law behaviour for the sequences studied regardless of the percent of *intron* contents. Havlin *et al.* (1995) state that DNA sequence in genes containing non-coding regions is correlated, and that the correlation is remarkably long range indeed, base pairs thousands of base pairs distant are correlated. Such LRC are not found in the coding regions of the gene. Havlin *et al.* (1995) suggest that non-coding regions in plants and invertebrates may display a smaller entropy and larger redundancy than coding regions, further supporting the possibility that non-coding regions of DNA may carry biological information. Investigations based on different models seem to suggest different results, as they all look into only a certain aspect of the entire DNA sequence. It is therefore important to investigate the degree of correlations in a model-independent way. Hence, one may ignore the composition of the four kinds of bases in coding and non-coding segments and only consider the rough structure of the complete genome or long DNA sequences. Yu *et al.* (2000) proposed a time series model based on the global structure of the complete genome and considered three kinds of length sequences. The values of the exponents from these three kinds of length sequences of bacteria indicate that the LRC exist in most of these sequences (Yu *et al.*, 2000 and all the references contained therein). Recently from a systematic analysis of human *exons*, coding sequences (CDS) and *introns*, Audit *et al.* (2001) have found that power law correlations (PLC) are not only present in non-coding sequences but also in coding regions somehow hidden in their inner codon structure. If it is now well admitted that LRC do exist in genomic sequence, their biological interpretation is still a continuing debate (Audit *et al.*, 2001 and all references therein).

The LRC does not necessarily imply a deviation from Gaussianity. For example, the fractional Brownian motion, which has Gaussian statistics, shows an inverse power-law spectrum. According to Allegrini *et al.* (1996, based on Levy's statistics), LRC would imply a strong deviation from Gaussian statistics while the investigation of Arneodo *et al.* (1995) yields an important conclusion that the DNA statistics are essentially Gaussian (Mohanty and Narayana Rao, 2000).

In visualizing very long DNA sequences, including the complete genomes of several bacteria, yeast and segments of human genes, it is seen that *fractal*-like patterns underlie these biological objects of prominent importance. The method used to visualize genomes of organisms may well be used as a convenient tool to trace, e.g., evolutionary relatedness of species (Hao *et al.*, 2000). Stanley *et al.* (1996) and Stanley *et al.* (1996c) discuss examples of complex systems composed of many interacting subsystems, which display non-trivial LRC or long-term "memory". The statistical properties of DNA sequences, heartbeat intervals, brain plaque in Alzheimer brains, and fluctuations in economics have the common feature that the guiding principle of scale invariance and universality appear to be relevant (Stanley, 2000).

### 6.1.3 *Nonlinear dynamics and chaos*

Irregular (nonlinear) fluctuations on all scales of space and time are generic to dynamical systems in nature such as fluid flows, atmospheric weather patterns, heart beat patterns, stock market fluctuations, etc. Mandelbrot (1977) coined the name *fractal* for the non-Euclidean geometry of such fluctuations which have fractional dimension, for example, the rise and subsequent fall with time of the Dow Jones Index or rainfall traces a zig-zag line in a two-dimensional plane and therefore has a *fractal* dimension greater than one but less than two. Mathematical models of dynamical systems are nonlinear and finite precision computer realizations exhibit sensitive dependence on initial conditions resulting in chaotic solutions, identified as deterministic chaos. *Nonlinear dynamics and chaos* is now (since 1980s) an area of intensive research in all branches of science (Gleick, 1987). The *fractal* fluctuations exhibit scale invariance or self-similarity manifested as the widely documented (Bak *et al.*, 1988; Bak and Chen, 1989, 1991; Schroeder, 1991; Stanley, 1995; Buchanan, 1997) inverse power-law form for power spectra of space-time fluctuations identified as *self-organized criticality* by Bak *et al.* (1987). The power-law is a distinctive experimental signature seen in a wide variety of complex systems. In economy it goes by the name fat tails, in physics it is referred to as critical fluctuations, in computer science and biology it is the edge of chaos, and in demographics it is called *Zipf's* law (Newman, 2000). Power-law scaling is not new to economics. The power-law distribution of wealth discovered by *Vilfredo Pareto* (1848–1923) in the 19th century (Eatwell *et al.*, 1991) predates any power-laws in physics (Farmer, 1999). One of the oldest scaling laws in geophysics is the *Omori law* (Omori, 1895). It describes the temporal distribution of the number of aftershocks, which occur after a larger earthquake (i.e., mainshock) by a scaling relationship. The other basic empirical seismological law, the *Gutenberg–Richter law* (Gutenberg and Richter, 1944) is also a scaling relationship, and relates intensity to its probability of occurrence (Hooge *et al.*, 1994). Time series analyses of global market economy also exhibits power-law behaviour (Bak *et al.*, 1992; Mantegna and Stanley, 1995; Sornette *et al.*, 1995; Chen, 1996a, 1996b; Stanley *et al.*, 1996a; Feigenbaum and Freund, 1997a, 1997b; Gopikrishnan *et al.*, 1999; Plerou *et al.*, 1999; Stanley *et al.*, 2000; Feigenbaum, 2001a, 2001b) with possible *multifractal* structure (Farmer, 1999) and has suggested an analogy to fluid turbulence (Ghashghaie *et al.*, 1996; Arneodo *et al.*, 1998). Sornette *et al.* (1995) conclude that the observed power law represents structures similar to *Elliott waves* of technical analysis first introduced in the 1930s. It describes the time series of a stock price as made of different waves; these waves are in relation to each other through the *Fibonacci* series. Sornette *et al.* (1995) speculate that *Elliott waves* could be a signature of an underlying critical structure of the stock market. Incidentally, the *Fibonacci* series represent a *fractal* tree-like branching network of self-similar structures (Stewart, 1992). The commonly found shapes in nature are the helix and the dodecahedron (Muller and Beugholt, 1996), which are signatures of self-similarity underlying *Fibonacci* numbers. The general systems theory (Chapter 2) shows that *Fibonacci* series underlies *fractal* fluctuations on all space-time scales.

Historically, basic similarity in the branching (*fractal*) form underlying the individual leaf and the tree as a whole was identified more than three centuries ago in botany (Arber, 1950). The branching (bifurcating) structure of roots, shoots, veins on leaves of plants, etc., have similarity in form to branched lighting strokes, tributaries of rivers, physiological networks of blood vessels, nerves and ducts in lungs, heart, liver, kidney, brain, etc. (Freeman, 1987, 1990; Goldberger *et al.*, 1990; Jean, 1994). Such seemingly complex network structure is again associated with *Fibonacci* numbers seen in the exquisitely ordered beautiful patterns in flowers and arrangement of leaves in the plant kingdom (Jean, 1994; Stewart, 1995). The identification of physical mechanism for the spontaneous generation of mathematically precise, robust spatial pattern formation in plants will have direct applications in all other areas of science (Mary Selvam, 1998). The importance of scaling concepts were recognized nearly a century ago in biology and botany where the dependence of a property $y$ on size $x$ is usually expressed by the allometric equation $y = ax^b$ where $a$ and $b$ are constants (Thompson, 1963; Strathmann, 1990;

Jean, 1994; Stanley, 1996b). This type of scaling implies a hierarchy of substructures and was used by *D'Arcy Thompson* for scaling anatomical structures, for example, how proportions tend to vary as an animal grows in size (West, 1990a). *D'Arcy Thompson* (1963, first published in 1917) in his book *On Growth and Form* has dealt extensively with similitude principle for biological modelling. Rapid advances have been made in recent years in the fields of biology and medicine in the application of scaling (*fractal*) concepts for description and quantification of physiological systems and their functions (Goldberger *et al.*, 1990; West, 1990a, 1990b; Deering and West, 1992; Skinner, 1994; Stanley *et. al.*, 1996b). In meteorological theory, the concept of self-similar fluctuations was identified and introduced in the description of turbulent flows by Richardson (1965, originally published in 1922; see also Richardson, 1960), Kolmogorov (1941, 1962), Mandelbrot (1975), Kadanoff (1996) and others (see Monin and Yaglom, 1975 for a review).

*Self-organized criticality* implies long-range space-time correlations or non-local connections in the spatially extended dynamical system. The physics underlying *self-organized criticality* is not yet identified. Prediction of the future evolution of the dynamical system requires precise quantification of the observed *self-organized criticality*. The author has developed a general systems theory (Capra, 1996), which predicts the observed *self-organized criticality* as a signature of quantum-like chaos in the macroscale dynamical system (Mary Selvam, 1990; Mary Selvam *et al.*, 1992; Selvam and Fadnavis, 1998). The model also provides universal and unique quantification for the observed *self-organized criticality* in terms of the statistical normal distribution.

Continuous periodogram power spectral analyses of the frequency distribution of bases A, C, G, T in Drosophila DNA base sequence agree with model prediction, namely, the power spectra follow the universal inverse power law form of the statistical normal distribution. The geometrical distribution in space, of the DNA bases, therefore exhibit *self-organized criticality*, which is a signature of quantum-like chaos. Earlier studies by the author have identified quantum-like chaos exhibited by dynamical systems underlying the observed *fractal* fluctuations of the following data sets: (1) Time series of meteorological parameters (Mary Selvam *et al.*, 1992; Selvam and Joshi, 1995; Selvam *et al.*, 1996;

Selvam and Fadnavis, 1998). (2) Spacing intervals of adjacent prime numbers (Selvam and Fadnavis, 1998; Selvam, 2001a). (3) Spacing intervals of adjacent non-trivial zeros of the Riemann zeta function (Selvam, 2001b).

## 6.2 General Systems Theory for Universal Quantification of Fractal Fluctuations of Dynamical Systems

As mentioned earlier (Section 1.3) power spectral analyses of *fractal* space-time fluctuations of dynamical systems exhibit inverse power-law form, i.e., a self-similar eddy continuum. The cell dynamical system model (Mary Selvam, 1990; Selvam and Fadnavis, 1998, and all references contained therein; Selvam, 2001a, 2001b) is a general systems theory (Capra, 1996) applicable to dynamical systems of all size scales. The model shows that such an eddy continuum can be visualized as a hierarchy of successively larger-scale eddies enclosing smaller-scale eddies. Eddy or wave is characterized by circulation speed and radius. Large eddies of root mean square (r.m.s) circulation speed $W$ and radius $R$ form as envelopes enclosing small eddies of r.m.s circulation speed $w_*$ and radius $r$ such that

$$W^2 = \frac{2}{\pi} \frac{r}{R} w_*^2 \qquad (6.1)$$

Since the large eddy is but the average of the enclosed smaller eddies, the eddy energy spectrum follows the statistical normal distribution according to the *Central Limit Theorem* (Ruhla, 1992). Therefore, the variance represents the probability densities. Such a result that the additive amplitudes of eddies, when squared, represent the probabilities is an observed feature of the subatomic dynamics of quantum systems such as the electron or photon (Maddox, 1988a, 1993; Rae, 1988). The *fractal* space-time fluctuations exhibited by dynamical systems are signatures of quantum-like mechanics. The cell dynamical system model provides a unique quantification for the apparently chaotic or unpredictable nature of such *fractal* fluctuations (Selvam and Fadnavis, 1998). The model predictions for quantum-like chaos of dynamical systems are as follows.

- The observed *fractal* fluctuations of dynamical systems are generated by an overall logarithmic spiral trajectory with the quasiperiodic *Penrose* tiling pattern (Nelson, 1986; Selvam and Fadnavis, 1998) for the internal structure.
- Conventional continuous periodogram power spectral analyses of such spiral trajectories will reveal a continuum of periodicities with progressive increase in phase.
- The broadband power spectrum will have embedded dominant wavebands, the bandwidth increasing with period length. The peak periods (or length scales) $E_n$ in the dominant wavebands is given in terms of $\tau$, the *golden mean* equal to $(1+\sqrt{5})/2$ ($\cong 1.618$) and $T_s$, the primary perturbation length scale as

$$E_n = T_s(2 + \tau)\tau^n \qquad (6.2)$$

Considering the most representative example of turbulent fluid flows, namely, atmospheric flows, Ghil (1994) reports that the most striking feature in climate variability on all time scales is the presence of sharp peaks superimposed on a continuous background. The model predicted periodicities (or length scales) in terms of the primary perturbation length scale units are 2.2, 3.6, 5.8, 9.5, 15.3, 24.8, 40.1, 64.9, 105.0, respectively, for values of $n$ ranging from $-1$ to 7. Periodicities (or length scales) close to model predicted have been reported in weather and climate variability (Burroughs, 1992; Kane, 1996), prime number distribution (Selvam, 2001a), Riemann zeta zeros (non-trivial) distribution (Selvam, 2001b).

As mentioned earlier Sornette *et al.* (1995) also conclude that the observed power-law represents structures similar to *Elliott waves* of technical analysis first introduced in the 1930s. It describes the time series of a stock price as made of different waves; these waves are in relation to each other through the *Fibonacci* series.

- The length scale ratio $r/R$ also represents the increment $d\theta$ in phase angle $\theta$ (Eq. (6.1)). Therefore, the phase angle $\theta$ represents the variance.

Hence, when the logarithmic spiral is resolved as an eddy continuum in conventional spectral analysis, the increment in wavelength is concomitant with increase in phase (Selvam and Fadnavis, 1998). Such a result that increments in wavelength and phase angle are related is observed in quantum systems and has been named *Berry's phase* (Berry, 1988; Maddox, 1988b; Simon *et al.*, 1988; Anandan, 1992). The relationship of angular turning of the spiral to intensity of fluctuations is seen in the tight coiling of the hurricane spiral cloud systems.

- The overall logarithmic spiral flow structure is given by the relation

$$W = \frac{w_*}{k} \log z \qquad (6.3)$$

- The constant $k$ in Eq. (6.3) is the steady state fractional volume dilution of large eddy by inherent turbulent eddy fluctuations. The constant $k$ is equal to $1/\tau^2 \cong 0.382$ and is identified as the universal constant for deterministic chaos in fluid flows (Selvam and Fadnavis, 1998). The steady state emergence of *fractal* structures is therefore equal to

$$\frac{1}{k} \cong 2.62 \qquad (6.4)$$

The model predicted logarithmic wind profile relationship such as Eq. (6.3) is a long-established (observational) feature of atmospheric flows in the atmospheric boundary layer, the constant $k$, called the *Von Karman's* constant has the value equal to 0.38 as determined from observations (Wallace and Hobbs, 1977).

- In Eq. (6.3), $W$ represents the standard deviation of eddy fluctuations, since $W$ is computed as the instantaneous r.m.s. eddy perturbation amplitude with reference to the earlier step of eddy growth. For two successive stages of eddy growth starting from primary perturbation $w_*$ the ratio of the standard deviations $W_{n+1}$ and $W_n$ is given from Eq. (6.3) as $(n + 1)/n$. Denoting by $\sigma$ the standard deviation of eddy fluctuations at the reference level ($n = 1$), the standard deviations of eddy fluctuations for successive stages of eddy growth are given as integer multiple of $\sigma$, i.e., $\sigma$, $2\sigma$, $3\sigma$, etc., and correspond respectively to statistical normalized standard deviation $t = 0, 1, 2, 3$, etc.

*Statistical normalized standard deviation t* = 0, 1, 2, 3,....... (6.5)

- The conventional power spectrum plotted as the variance versus the frequency in log–log scale will now represent the eddy probability density on logarithmic scale versus the standard deviation of the eddy fluctuations on linear scale since the logarithm of the eddy wavelength represents the standard deviation, i.e., the r.m.s. value of eddy fluctuations (Eq. (6.3)). The r.m.s. value of eddy fluctuations can be represented in terms of statistical normal distribution as follows. A normalized standard deviation $t = 0$ corresponds to cumulative percentage probability density equal to 50 for the mean value of the distribution. Since the logarithm of the wavelength represents the r.m.s. value of eddy fluctuations the normalized standard deviation $t$ is defined for the eddy energy as

$$t = \frac{\log L}{\log T_{50}} - 1 \qquad (6.6)$$

- The parameter $L$ in Eq. (6.6) is the wavelength (or period) and $T_{50}$ is the wavelength (or period) up to which the cumulative percentage contribution to total variance is equal to 50 and $t = 0$. The variable $\log T_{50}$ also represents the mean value for the r.m.s. eddy fluctuations and is consistent with the concept of the mean level represented by r.m.s. eddy fluctuations. Spectra of time series of fluctuations of dynamical systems, for example, meteorological parameters, when plotted as cumulative percentage contribution to total variance versus $t$ follow the model predicted universal spectrum (Selvam and Fadnavis, 1998, and all references therein). The literature shows many examples of pressure, wind and temperature whose shapes display a remarkable degree of universality (Canavero and Einaudi, 1987).
- The periodicities (or length scales) $T_{50}$ and $T_{95}$ up to which the cumulative percentage contribution to total variances are respectively equal to 50 and 95 are computed from model concepts as follows. The power spectrum, when plotted as normalized standard deviation $t$ versus cumulative percentage contribution to total variance represents the statistical normal distribution (Eq. (6.6)), i.e., the variance represents

the probability density. The normalized standard deviation values $t$ corresponding to cumulative percentage probability densities $P$ equal to 50 and 95, respectively, are equal to 0 and 2 from statistical normal distribution characteristics. Since $t$ represents the eddy growth step $n$ (Eq. (6.5)) the dominant periodicities (or length scales) $T_{50}$ and $T_{95}$ up to which the cumulative percentage contribution to total variance are respectively equal to 50 and 95 are obtained from Eq. (6.2) for corresponding values of $n$ equal to 0 and 2. In the present study of *fractal* fluctuations of frequency distribution of Drosophila DNA bases A, C, G, T, the primary perturbation length scale $T_s$ is equal to unit length segment of 50 bases and $T_{50}$ and $T_{95}$ are obtained as

$$T_{50} = (2 + \tau)\tau^0 \cong 3.6 \; unti \; length \; segment \; of \; 50 \; bases \qquad (6.7)$$

$$T_{95} = (2 + \tau)\tau^2 \cong 9.5 \; unti \; length \; segment \; of \; 50 \; bases \qquad (6.8)$$

The aforementioned model predictions are applicable to all real world and computed model dynamical systems. Continuous periodogram power spectral analyses of number frequency (per 50 bases) of occurrence of bases A, C, G, T in Drosophila DNA base sequence at different locations along its length give results in agreement with the aforesaid model predictions.

## 6.3 Data and Analysis

The Drosophila DNA base sequence was obtained from Berkeley Drosophila Genome Project (BGDP Resources at http://www.fruitfly. org/index.html. The data set used for the study corresponds to the file NA_ARMS~1 with the title: >2L, 28-11-2001.1 (22207800 bases) segment 1 of 1 for arm 2L on wed Nov 28 00:30:01 PST 2001 (https:// www.fruitfly.org/sequence/dlMfasta.shtml) finished sequence for 2L. The first 225,000 bases were used to give 50 data sets each of length 4,500 bases. The number of times that each of the bases A, C, G, T occur in successive blocks of 50 bases was determined for each data set of 4,500 bases. Each data set of 4,500 bases then gives 4 groups of 90 frequency sequence values corresponding respectively to the four bases A, C, G, T.

### 6.3.1  *Fractal nature of frequency distribution of Drosophila DNA base (A, C, G or T) sequence*

A representative sample for the frequency of occurrence of base A in successive blocks of length 50 bases is plotted in Fig. 6.1 for 10, 100, 1,000

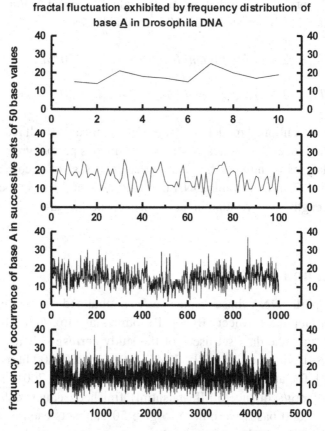

**Figure 6.1:**    Fractal nature of frequency distribution of Drosophila DNA base (A, C, G or T) sequence.

and 4,500 segments for the total sequence consisting of 225,000 bases used in the study.

The frequency distribution shows irregular or *fractal* fluctuations for all the segment length scales. The irregular fluctuations may be visualized to result from the superimposition of an ensemble of eddies (wavelengths).

## 6.3.2 *The frequency distributions of DNA bases A, C, G, T and the statistical normal distribution*

The frequency distribution of bases A, C, G, T follow statistical normal distribution as described in the following. Each data set consists of the frequency distribution $X_j$ where $j = 1, 2, \dots n$ denotes the class interval number, the total number $n$ equals 90 class intervals and each class interval consists of 50 bases, so that each data set consists of 4,500 bases. The mean *Xbar*, *standard deviations, and normalized standard deviation* $t_j$ for each set of frequency distributions was calculated as follows:

$$mean\ Xbar = \frac{\sum_{1}^{n} X_j \times j}{n}$$

$$standard\ deviation\ s = \sqrt{\frac{\sum_{1}^{n} (mean - X_j)^2 \times j}{n}}$$

The cumulative frequency of occurrence $p_j$ of base (A, C, G or T) for class intervals $j = 1, 2, \dots n$ were calculated as

$$p_j = \sum_{1}^{j} X_j$$

The cumulative percentage frequency of occurrence $p_c$ of base (A, C, G or T) for class intervals $j = 1, 2, \dots n$ were then calculated as

$$p_c = \frac{p_j}{p_n} \times 100$$

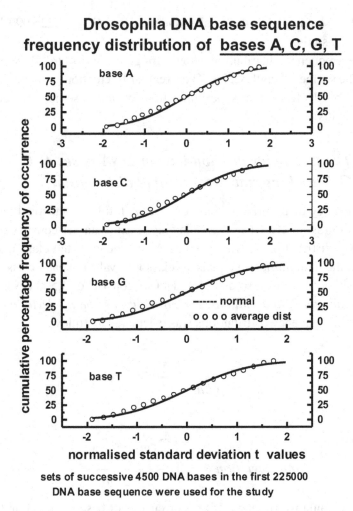

**Drosophila DNA base sequence frequency distribution of bases A, C, G, T**

normalised standard deviation t values

sets of successive 4500 DNA bases in the first 225000
DNA base sequence were used for the study

**Figure 6.2:**   The cumulative percentage frequency of occurrence of bases A, C, G, T in Drosophila DNA sequence follow closely the statistical normal distribution.

The graph of cumulative percentage frequency of occurrence $p_c$ versus the corresponding normalized standard deviation $t_j$ follows closely the statistical normal distribution as shown in Fig. 6.2 for all the four bases A, C, G, T in the Drosophila DNA sequence. The aforementioned result is consistent with model prediction that the variance spectrum of *fractal* fluctuations follows statistical normal distribution as explained in the following. From Eq. (6.1), namely

$$W^2 = \frac{2}{\pi}\frac{r}{R}w_*^2$$

it is seen that the length scale ratio $r/R$ (or frequency ratio) represents the variance spectrum ($W^2/w_*^2$) and therefore the cumulative frequency distribution follows closely the cumulative normal distribution as shown in Fig. 6.2.

### 6.3.3 *Continuous periodogram power spectral analyses*

The broadband power spectrum of space-time fluctuations of dynamical systems can be computed accurately by an elementary, but very powerful method of analysis developed by Jenkinson (1977) which provides a quasicontinuous form of the classical periodogram allowing systematic allocation of the total variance and degrees of freedom of the data series to logarithmically spaced elements of the frequency range (0.5, 0). The periodogram is constructed for a fixed set of 10,000($m$) wavelengths (or periodicities) $L_m$ which increase geometrically as $L_m = 2exp(Cm)$ where $C = 0.001$ and $m = 0, 1, 2, ... m$. The data series $X_j$ for the $n$ data points was used. The periodogram estimates the set of $A_m \cos(2\pi v_m S - \phi_m)$ where $A_m$, $v_m$ and $\phi_m$ denote respectively the amplitude, frequency and phase angle for the $m$th wavelength (or periodicity) and $S$ is the spatial (or time) interval in units of 50 bases in the present study of Drosophila DNA base sequence structure. The cumulative percentage contribution to total variance was computed starting from the high frequency side of the spectrum. The wavelength (or period) $T_{50}$ at which 50% contribution to total variance occurs is taken as reference and the normalized standard deviation $t_m$ values are computed as (Eq. (6.6)).

$$t_m = \frac{\log L_m}{\log T_{50}} - 1$$

The cumulative percentage contribution to total variance, the cumulative percentage normalized phase (normalized with respect to the total phase rotation) and the corresponding $t_m$ values were computed. The power spectra were plotted as cumulative percentage contribution to total variance versus the normalized standard deviation $t_m$ as given above. The wavelength (or period) $L_m$ is in units of 50 bases as explained above. Wavelengths (or periodicities) up to $T_{50}$ contribute up to 50% of total variance. The phase spectra were plotted as cumulative percentage normalized (normalized to total rotation) phase

## 6.3.4 *Power spectral analyses: Summary of results*

### 6.3.4.1 *Average variance and phase spectra*

The average variance and phase spectra for the 50 data sets used in the study along with statistical normal distribution are shown in Fig. 6.3 for the four bases A, C, G, T. The "goodness of fit" (statistical chi-square test) between the variance spectra and statistical normal distribution is significant at less than or equal to 5% level for all the variance spectra. The eddy variance spectra following statistical normal distribution is a signature of quantum-like chaos (see Chapter 2) in the frequency distribution sequence of bases A, C, G, T in Drosophila DNA base sequence arrangement. Phase spectra are close to the statistical normal distribution, with the "goodness of fit" being statistically significant for 42%, 36%, 48% and 42% of data sets, respectively, for the four bases A, C, G, T.

However, in all the cases, the "goodness of fit" between variance and phase spectra is statistically significant (chi-square test) for individual dominant wavebands, in particular for shorter wavelengths as shown in Fig. 6.6. Eddy variance spectra following phase spectra is identified as *Berry's phase* and is also a signature of quantum-like chaos (see Chapter 1, Eq. (1.1)). The data sets, which do not exhibit *Berry's phase*, are indicated in Fig. 6.9.

### 6.3.4.2 *Dominant wavebands*

The power spectra exhibit dominant wavebands where the normalized variance is equal to or greater than 1. The dominant peak wavelengths (periodicities) were grouped into class intervals 2–3, 3–4, 4–6, 6–12, 12–20, 20–30, 30–50, 50–80, 80–120. These class intervals include the model predicted (Eq. (6.2)) dominant peak periodicities (or length scales)

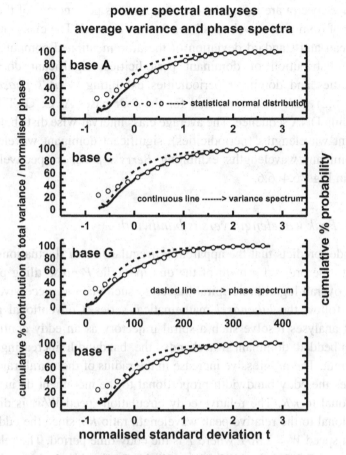

**power spectral analyses**
**average variance and phase spectra**

**Figure 6.3:** Average variance (continuous line) and phase (dashed line) spectra for the bases A, C, G, T for the 50 data sets used in the study. The statistical normal distribution (open circles) is also shown.

2.2, 3.6, 5.8, 9.5, 15.3, 24.8, 40.1, 64.9, 105.0 (in block length segment unit of 50 bases) for values of *n* ranging from −1 to 7. Wavelength class interval-wise percentage frequency of occurrence of dominant periodicities was computed. In each class interval, the number of dominant statistically significant (less than or equal to 5%) periodicities and also the number of dominant wavebands which exhibit *Berry's phase* (variance and phase spectra are the same) are computed as percentages of the total number of dominant wavebands in each class interval. The class interval-wise mean and standard deviation of the aforementioned computed frequency distribution of dominant periodicities, significant dominant periodicities and dominant periodicities exhibiting *Berry's phase* (see Section 6.2) were then computed for the four bases A, C, G, T in the Drosophila DNA sequence. The average class interval-wise distribution of dominant wavelengths (periodicities), significant dominant wavelengths and dominant wavelengths exhibiting *Berry's phase* respectively are shown in Figs. 6.4–6.6.

### 6.3.4.3 *Peak wavelength versus bandwidth*

The model predicts that the apparently irregular *fractal* fluctuations contribute to the ordered growth of the quasiperiodic *Penrose* tiling pattern with an overall logarithmic spiral trajectory such that the successive radii lengths follow the *Fibonacci* mathematical series. Conventional power spectral analyses resolve such a spiral trajectory as an eddy continuum with embedded dominant wavebands, the bandwidth increasing with wavelength. The progressive increase in the radius of the spiral trajectory generates the eddy bandwidth proportional to the increment $d\theta$ in phase angle equal to $r/R$. The relative eddy circulation speed $W/w_*$ is directly proportional to the relative peak wavelength ratio $R/r$ since the eddy circulation speed $W = 2\pi R/T$ where $T$ is the eddy time period. The relationship between the peak wavelength and the bandwidth is obtained from Eq. (6.1), namely

$$W^2 = \frac{2}{\pi}\frac{r}{R}w_*^2$$

Considering eddy growth with overall logarithmic spiral trajectory

$$relative\ eddy\ bandwidth \propto d\theta \propto \frac{r}{R}$$

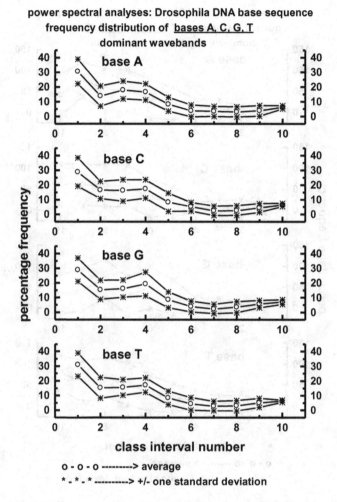

**Figure 6.4:**  Average wavelength class interval-wise distribution of dominant wavebands for the four bases A, C, G, T in the 50 data sets (a total of 225,000 bases) of Drosophila DNA.

The eddy circulation speed is related to eddy radius as

$$W = \frac{2\pi R}{T}$$

$$W \propto R \propto peak\ wavelength$$

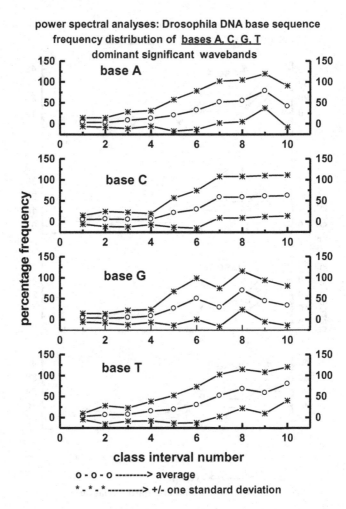

power spectral analyses: Drosophila DNA base sequence
frequency distribution of <u>bases A, C, G, T</u>
dominant significant wavebands

o - o - o --------> average

* - * - * ----------> +/- one standard deviation

**Figure 6.5:**    Average wavelength class interval-wise distribution of dominant significant wavebands for the four bases A, C, G, T in the 50 data sets (a total of 225,000 bases) of Drosophila DNA base sequence used for the study.

The relative peak wavelength is given in terms of eddy circulation speed as

$$relative\ peak\ wavelength \propto \frac{W}{w_*}$$

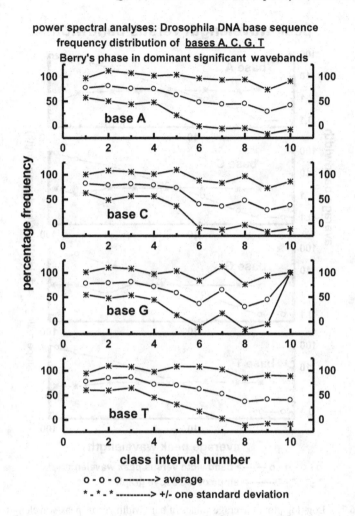

**Figure 6.6:** Average wavelength class interval-wise distribution of dominant wavebands exhibiting Berry's phase for the four bases A, C, G, T in the 50 data sets (a total of 225,000 bases) of Drosophila DNA base sequence used for the study.

From Eq. (6.1) the relationship between eddy bandwidth and peak wavelength is obtained as

$$eddy\ bandwidth = (peak\ wavelength)^2$$

$$\frac{\log(eddy\ bandwidth)}{\log(peak\ wavelength)} = 2$$

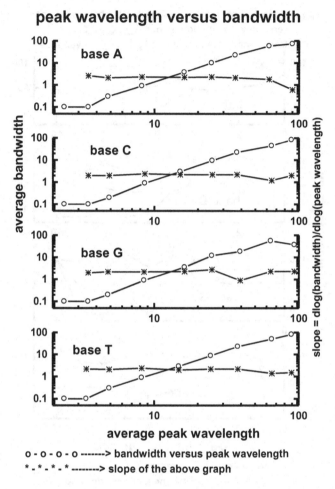

**peak wavelength versus bandwidth**

**Figure 6.7:**    Log–log plot of average values of bandwidth versus peak wavelength for the four bases A, C, G, T. The slope (bandwidth/peak wavelength) of this graph, also plotted in the figure shows an approximately constant value equal to about 2.

A log–log plot of peak wavelength versus bandwidth will be a straight line with a slope (bandwidth/peak wavelength) equal to 2. A log–log plot of the average values of bandwidth versus peak wavelength shown in Fig. 6.7 exhibits a constant slope approximately equal to 2 in agreement with the aforementioned model prediction.

The mean and standard deviation of the frequency distribution for bases A, C, G, T for all the 50 data sets are given in Fig. 6.8. Each data set

**mean and standard deviation of frequency distribution of Drosophila DNA bases A, C, G, T**

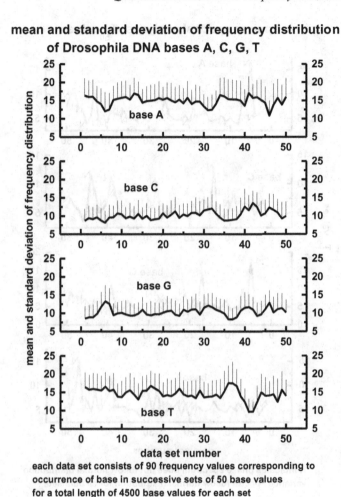

each data set consists of 90 frequency values corresponding to
occurrence of base in successive sets of 50 base values
for a total length of 4500 base values for each set

**Figure 6.8:** The mean and standard deviation of the frequency distribution for bases A, C, G, T for all the 50 data sets. The vertical line corresponds to one standard deviation for the data set.

consists of a sequence of 90 frequency values corresponding to 90 successive block lengths of 50 bases of Drosophila DNA base sequence.

The periodicities $T_{50}$ up to which the cumulative percentage contribution to total variance is equal to 50 are shown for the bases A, C, G, T for the 50 data sets in Fig. 6.9. The letter "N" denotes data set which does not exhibit *Berry's phase*, i.e., the "goodness of fit" between variance and phase spectra is not significant.

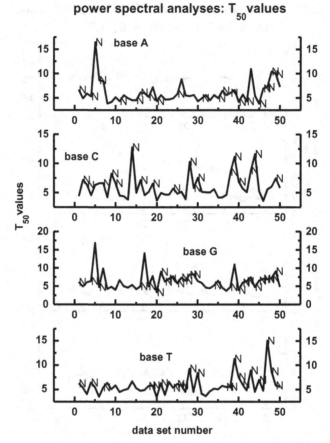

**power spectral analyses: $T_{50}$ values**

N indicates phase spectrum does not follow normal distribution
variance spectra follow normal distribution for all data sets

**Figure 6.9:** The periodicities $T_{50}$ up to which the cumulative percentage contribution to total variance is equal to 50. The letter "N" denotes data set which does not exhibit *"Berry's phase"*, i.e., the "goodness of fit" between variance and phase spectra is not significant. Variance spectra follow normal distribution for all data sets.

## 6.4 Results and Discussion

The number frequency of occurrence of each of the bases A, C, G, T in successive block lengths of 50 bases of Drosophila DNA base sequence exhibit self-similar *fractal* fluctuations generic to dynamical systems in nature. The apparently irregular (chaotic) *fractal* fluctuations, which characterize the fine-scale geometry of spatial structures in nature, are now an intensive field of study in the new science of *Nonlinear Dynamics and Chaos*. The *fractal* fluctuations are basically a zigzag pattern of successive upward and downward swings such as that shown in Fig. 6.1 for the frequency distribution of bases A, C, G, T for all data lengths, i.e., number of blocks ranging from 1 to the maximum 4,500, a total of 225,000 Drosophila DNA base sequence. Such irregular fluctuations may be visualized to result from the superimposition of a continuum of eddies. Power spectral analysis is commonly applied to resolve the component wavelengths and their phases, the wavelengths being given in terms of the unit block length of 50 bases used for determining the wavelength distribution. Continuous periodogram power spectral analyses of the *fractal* fluctuations in the frequency distribution of bases A, C, G, T in Drosophila DNA base sequence follow closely the following model predictions given in Section 1.2.

- The variance spectra for the entire data sets exhibit the universal inverse power-law form $1/f^{\alpha}$ of the statistical normal distribution (Figs. 6.2 and 6.3) where $f$ is the frequency and $\alpha$, the spectral slope decreases with increase in wavelength (or decrease in frequency since frequency is inversely proportional to wavelength) and approaches 1 for long wavelengths. Inverse power-law form for power spectra imply long-range spatial correlations in the frequency distributions of the bases A, C, G, T in Drosophila DNA base sequence structure. *Fractal* fluctuations exhibit scale invariance, namely the eddy amplitudes being related to each other by a simple proportionality factor for the range of wavelengths for which $\alpha$ is a constant. The observed frequency distribution exhibits *multifractal* structure since the slope $\alpha$ of the spectrum is not a constant, but decreases with increasing wavelength. Microscopic-scale quantum systems such as the electron or photon exhibit non-local connections or LRC and are visualized to result from the superimposition of a continuum of eddies. Therefore, by analogy, the observed *fractal* fluctuations of frequency distribution of bases A, C, G, T exhibit quantum-like chaos in the Drosophila DNA base sequence structure.

Incidentally physics at the atomic scale is determined by the rules of quantum mechanics, which tells us that particles propagate like waves, and so can be described by a quantum mechanical wave function (Rae, 1999). As an immediate consequence, a particle can be in two or more states at the same time — a so-called superposition of states. This curious behaviour has been hugely successful in describing physical systems at the microscopic level. For example, under the rules of quantum mechanics two atoms sharing an electron form a chemical bond, whereas in classical theory the electron remains confined to one atom and the bond cannot form (Blatter, 2000).

- *Berry's phase*, namely, phase spectra and variance spectra being the same is seen in about 50% of the data sets (Fig. 6.9). However, for all the data sets, the phase spectra follow the variance spectra for a majority of dominant wavebands (Fig. 6.6), particularly for the shorter wavelengths up to 4–6 unit block length of 50 bases. Microscopic scale quantum systems exhibit *Berry's phase*.
- The period $T_{50}$ up to which the cumulative percentage contribution to total variance is equal to 50% is larger than the model predicted (Eq. (6.7)) value equal to 3.6 unit block length of 50 bases for a majority of data sets (Fig. 6.9). This may indicate that the primary length scale may be less than the unit block length of 50 bases used for evaluating the frequency distribution.
- The power spectra exhibit dominant wavebands with peak wavelengths close to model predicted values (Eq. (6.2)). The average class interval-wise distribution of dominant wavelengths (Fig. 6.4) and dominant wavelengths which exhibit *Berry's phase* (Fig. 6.6) for all data sets show a maximum for the shorter wavelengths up to 4–6 unit block length of 50 bases. The dominant significant wavelengths show a maximum for wavelengths larger than 4–6 unit block length of 50 bases. This result is consistent with observed value of $T_{50}$ being greater than the model predicted value equal to 3.6 unit block length of 50 bases as shown the earlier item above.
- The bandwidth of the dominant waveband is directly proportional to the square of the corresponding peak wavelength (Fig. 6.7) in agreement with model prediction (Eq. (6.9)).

## 6.5 Conclusions

Power spectra of frequency distribution of bases A, C, G, T of Drosophila DNA base sequence follow the model predicted universal and unique inverse power-law form of the statistical normal distribution.

Inverse power-law form for power spectra generic to *fractal* fluctuations is a signature of *self-organized criticality* in dynamical systems in nature. The author had shown earlier (Selvam and Fadnavis, 1998; Selvam, 2001a, 2001b) that (a) *self-organized criticality* can be quantified in terms of the universal inverse power-law form of the statistical normal distribution and (b) *self-organized criticality* of self-similar *fractal* fluctuations implies long-range space-time correlations and is a signature of quantum-like chaos in macroscale dynamical systems of all space-time scales.

Inverse power-law form for power spectra of fluctuations in spatial distribution of bases A, C, G, T imply long-range spatial correlations, or in other words, persistence or long term (length scale) memory of short-term fluctuations. The fine-scale structure of longer length scale fluctuations carries the signature of shorter length scale fluctuations. The cumulative integration of shorter length scale fluctuations generates longer length scale fluctuations (eddy continuum) with two-way ordered energy feedback between the fluctuations of all length scales (Eq. (6.1)). The eddy continuum acts as a robust unified whole fuzzy logic network with global response to local perturbations. Increase in random noise or energy input into the short-length scale fluctuations creates intensification of fluctuations of all other length scales in the eddy continuum and may be noticed immediately in shorter length scale fluctuations. Noise is therefore a precursor to signal.

Real world examples of noise enhancing signal has been reported in electronic circuits (Brown, 1996). Man-made, urbanization related, greenhouse gas induced global warming (enhancement of small-scale fluctuations) is now held responsible for devastating anomalous changes in regional and global weather and climate in recent years (Selvam and Fadnavis, 1998). Noise and fluctuations are at the seat of all physical phenomena. It is well known that, in linear systems, noise plays a destructive role. However, an emerging paradigm for nonlinear systems is that noise can play a constructive role — in some cases information transfer can be optimized at non-zero noise levels. Another use of noise is that its measured characteristics can tell us useful information about the system itself. Problems associated with fluctuations have been studied since 1826 (Abbott, 2001).

The apparently irregular *fractal* fluctuations of the frequency distribution of bases A, C, G, T in Drosophila DNA base sequence self-organize spontaneously to generate the robust geometry of logarithmic spiral with the quasiperiodic *Penrose* tiling pattern for the internal structure. Conventional power spectral analyses resolves such a logarithmic spiral geometry as an eddy continuum with embedded dominant wavebands, the peak periodicities being functions of the *golden mean* and the primary perturbation length scale equal to block length of 50 bases used in the present study. Power spectral analyses of the frequency distribution of bases A, C, G, T in Drosophila DNA base sequence also exhibit the model predicted dominant wavebands. These dominant periodicities are intrinsic to the self-similar *fractal* fluctuations (space-time) of dynamical systems in nature. Quantum systems are also characterized by continuous irregular space-time fluctuations analogous to *fractal* fluctuations of macroscale dynamical systems (Hey and Walters, 1989).

The quasicrystalline structure of the quasiperiodic *Penrose* tiling pattern underlies the apparently irregular distribution of the bases A, C, G, T in Drosophila DNA base sequence. Historically, Schrodinger (1967) introduced a concept that the most essential part of a living cell — the chromosome fibre — maybe suitably called an aperiodic crystal (Gribbin, 1985). A periodic crystal, like one of common salt, can carry only a very limited amount of information. But an aperiodic crystal in which there is structure obeying certain fundamental laws, but no dull repetition can carry enormous amount of information (Gribbin, 1985). The space filling geometric figure of the *Penrose* tiling pattern has intrinsic local 5-fold symmetry (Devlin, 1997) and also 10-fold symmetry. One of the three basic components of DNA, the deoxyribose is a five-carbon sugar and may represent the local 5-fold symmetry of the quasicrystalline structure of the quasiperiodic *Penrose* tiling pattern of the DNA molecule as a whole. The DNA molecule also shows 10-fold symmetry in the arrangement of 10 bases per turn of the double helix (Watson and Crick, 1953). The study of plant *phyllotaxis* in botany shows that the quasicrystalline structure of the quasiperiodic *Penrose* tiling pattern provides maximum packing efficiency for seeds, florets, leaves, etc. (Jean, 1994; Stewart, 1995; Levine and Steinhardt, 1984; Mary Selvam, 1998). Quasicrystalline structure of the quasiperiodic *Penrose* tiling pattern may be the geometrical structure underlying the packing of $10^3$–$10^5$ micrometer of DNA in a eukaryotic (higher organism) chromosome into a metaphase structure a few microns long.

The important result of the present study is that the observed *fractal* frequency distributions of the bases A, C, G, T of Drosophila DNA base sequence exhibit long-range spatial correlations or *self-organized criticality* generic to dynamical systems in nature. Therefore, artificial modification of base sequence structure at any location may have significant noticeable effect on the function of the DNA molecule as a whole. Further, the presence of *introns* may not be redundant, but may serve to organize the effective functioning of the *exons* in the DNA molecule as a complete unit.

In summary, the precise geometrical pattern of the quasiperiodic Penrose tiling pattern underlies the apparently chaotic *fractal* frequency distribution of base sequence in Drosophila DNA. The spatial geometry of the DNA is therefore organized into a hierarchy of helical structures (vortices) such as those seen in turbulent fluid flows. Such a concept may explain the observed loops of DNA in metaphase chromosome (Grosveld and Fraser, 1997) and also the characteristic and reproducible banding pattern of polytene chromosome (Corces and Gerasimova, 1997). It is believed that the loop organization of chromatin is important not only for compaction and spatial organization of the chromatin but also for the regulation of gene expression. Each loop domain may represent an independent unit of chromatin structure and gene activity (Luderus and van Driel, 1997). Audit *et al.* (2002) discuss analyses of results (wavelet transform) with regard to possible interpretations of the observed LRC in terms of mechanisms that might govern the positioning and the dynamics of the nucleosome along the DNA chain through cooperative process. Shiba *et al.* (2002) assessed the significance of periodicities of DNA in the origin of genes by constructing such periodic DNAs. Herzel *et al.* (1999) show that correlations within proteins affect mainly the oscillations at distances below 35 bp. The long-ranging correlations up to 100 bp reflect primarily DNA folding. Since the topological state of genomic DNA is of importance for its replication, recombination and transcription, there is an immediate interest to obtain information about the supercoiled state from sequence periodicities (Herzel *et al.*, 1998, 1999). Fourier transform analysis applied to a DNA sequence offers a great new avenue for extracting information on the evolution of a DNA sequence (Nagai *et al.*, 2001). Ordered patterns organized in hierarchical periodicities were identified in DNA subtelomeric sequences from two lower eukaryotes, *P. falciparum* and *S. cerevisiae* (Pizzi *et al.*, 1990). Identification of dominant periodicities in DNA sequence will help understand the important role of coherent structures in genome sequence organization (Chechetkin and Turygin, 1995; Widom, 1996).

## Acknowledgement

The author is grateful to Dr. A. S. R. Murty for encouragement during the course of the study.

## References

Abbott, D. (2001). Overview: Unsolved problems of noise and fluctuations, *Chaos*, 11(3), pp. 526–538.

Alcamo, E. (2001). *DNA Technology*, 2nd Ed. (Academic Press, New York), p. 339.

Allegrini, P., Barbi, M., Grigolini, P., and West, B. J. (1996). Dynamical model for DNA sequences, *Phys. Rev. E*, 52(5), pp. 5281–5296. http://linkage.rockefeller.edu/wli/dna_corr.

Anandan, J. (1992). The geometric phase, *Nature*, 360, pp. 307–313.

Arber, A. (1950). *The Natural Philosophy of Plant Form* (University Press, London).

Arneodo, A., Bacry, E., Graves, P. V., and Muzy, J. F. (1995). Characterizing long-range correlations in DNA sequences from wavelet analysis, *Phys. Rev. Lett.*, 74(16), pp. 3293–3296. http://linkage.rockefeller.edu/wli/dna_corr/arneodo95.pdf.

Arneodo, A., Muzy, J.-F., and Sornette, D. (1998). "Direct" causal cascade in the stock market, *Eur. Phys. J. B*2, pp. 277–282.

Audit, B., Thermes, C., Vaillant, C., d'Aubenton-Carafa, Y., Muzy, J. F., and Arneodo, A. (2001). Long-range correlations in genomic DNA: A signature of the nucleosomal structure, *Phys. Rev. Lett.*, 86(11), pp. 2471–2474. http:// linkage.rockefeller.edu/wli/dna_corr/audit01.pdf.

Audit, B., Vaillant, C., Arneodo, A., d'Aubenton-Carafa, Y., and Thermes, C. (2002). Long-range correlations between DNA bending sites: Relation to the structure and dynamics of nucleosomes, *J. Mol. Biol.*, 316(4), pp. 903–918.

Bak, P., Tang, C., and Wiesenfeld, K. (1987). Self-organized criticality: An explanation of $1/f$ noise, *Phys. Rev. Lett.*, 59, 381–384.

Bak, P. C., Tang, C., and Wiesenfeld, K. (1988). Self-organized criticality, *Phys. Rev. A.*, 38, pp. 364–374.

Bak, P., Chen, K. (1989). The physics of fractals, *Physica D*, 38, pp. 5–12.

Bak, P., Chen, K. (1991). Self-organized criticality, *Sci. Am.*, Jan., pp. 26–33.

Bak, P., Chen, K., Scheinkman, J. A. and Woodford, M. (1992). *Self-Organized Criticality and Fluctuations in Economics* (Santa Fe Institute, Santa Fe). http://www.santafe.edu/sfi/publications/Abstracts/92-04-018abs.html.

Ball, P. (2000). Augmenting the alphabet, *Nature Science Update*, 30 August.

Bates, A. D. and Maxwell, A. (1993). *DNA Topology* (Oxford University Press, Oxford) p. 111.

Berry, M. V. (1988). The geometric phase, *Sci. Amer.*, Dec., pp. 26–32.

Blatter, G. (2000). Schrodinger's cat is now fat, *Nature*, 406, pp. 25–26.

Brown, J. (1996). Where two worlds meet, *New Scientist*, 18 May, pp. 26–30.

Buchanan, M. (1997). One law to rule them all, *New Scientist*, 8 November, pp. 30–35.

Buldyrev, S. V., Goldberger, A. L., Havlin, S., Mantegna, R. N., Matsa, M. E., Peng, C. K., Simons, M., and Stanley, H. E. (1995). Long-range correlation properties of coding and non-coding DNA sequences — GenBank analysis, *Phys. Rev. E*, 51(5), pp. 5084–5091. http://linkage.rockefeller.edu/wli/dna_ corr/buldyrev95.pdf.

Burroughs, W. J. (1992) *Weather Cycles: Real or Imaginary?* (Cambridge University Press, Cambridge).

Canavero, F. G. and Einaudi, F. (1987). Time and space variability of atmospheric processes, *J. Atmos. Sci.*, 44(12), pp. 1589–1604.

Capra, F. (1996). *The Web of Life* (Harper Collins, London), p. 311.

Chatzidimitriou-Dreismann, C. A., and Larhammar, D. (1993). Scientific Correspondence, *Nature*, 361, pp. 212–213. http://linkage.rockefeller.edu/ wli/dna_corr.

Chechetkin, V. R. and Yu. Turygin, A. (1995). Search of hidden periodicities in DNA sequences, *J. Theor. Biol.*, 175, pp. 477–494. http://linkage.rockefeller.edu/wli/dna_corr.

Chen, P. (1996a). Trends, shocks, persistent cycles in evolving economy — Business cycle measurement in time-frequency representation. In *Nonlinear Dynamics and Economics*, eds., Barnett, W. A., Kirman, A. P., and Salmon, M., Chapter 13 (Cambridge University Press, Cambridge).

Chen, P. (1996b). A random walk or color chaos on the stock market? Time-frequency analysis of S&P Indexes, *Stud. Nonlinear Dyn. Econom.*, 1(2), pp. 87–103. http://mitpress.mit.edu/e-journals/SNDE/001/articles/v1n2002.pdf.

Claire, J. (1964). *The Stuff of Life* (Phoenix House, London), p. 67.

Clark, A. G. (2001). The search for meaning in noncoding DNA, *Genome Res.*, 11, pp. 1319–1320. http://linkage.rockefeller.edu/wli/dna_corr.

Corces, V. G. and Gerasimova, T. I. (1997). Chromatin domains and boundary elements. In *Nuclear Organization, Chromatin Structure, and Gene Expression*, eds. Van Driel, R. and Otte, A. P. (Oxford University Press, Oxford), pp. 83–94.

Deering, W. and West, B. J. (1992). Fractal physiology, *IEEE Eng. Med. Biol.*, 11(2), pp. 40–46.

Devlin, K. (1997). *Mathematics: The Science of Patterns* (Scientific American Library, New York), p. 101.

Eatwell, J., Milgate, M., and Newman, P. (1991). *The New Palgrave: A Dictionary of Economics 3* (MacMillan Press, London).

Farmer, J. D. (1999). Physicists attempt to scale the ivory towers of finance, *Computing in Science & Engineering*, November/December, pp. 26–39. http://www.santafe.edu/sfi/publications/Abstracts/99-10-073abs.html.

Feigenbaum J. A. and Freund, P. G. O. (1997a). Discrete scaling in stock markets before crashes. http://xxx.lanl.gov/pdf/cond-mat/9509033.

Feigenbaum, J. A. and Freund, P. G. O. (1997b). Discrete scale invariance and the "second Black Monday". http://xxx.lanl.gov/pdf/cond-mat/9710324.

Feigenbaum, J. A. (2001a). A statistical analysis of log-periodic precursors to financial crashes. http://xxx.lanl.gov/pdf/cond-mat/0101031.

Feigenbaum, J. A. (2001b). More on a statistical analysis of log-periodic precursors to financial crashes. http://xxx.lanl.gov/pdf/cond-mat/0107445.

Freeman, G. R. (1987) Introduction. In *Kinetics of Nonhomogenous Processes*, ed. Freeman, G. R. (John Wiley and Sons, Inc., New York), pp. 1–18.

Freeman, G. R. (1990). *KNP89*: Kinetics of non-homogenous processes (*KNP*) and nonlinear dynamics, *Can. J. Phys.*, 68, pp. 655–659.

Ghashghaie, S., Breymann, P. J., Talkner, P., Dodge, Y. (1996). Turbulent cascades in foreign exchange markets, *Nature*, 381, pp. 767–770.

Ghil, M. (1994). Cryothermodynamics: The chaotic dynamics of paleoclimate, *Physica D*, 77, pp. 130–159.

Gleick, J. (1987) *Chaos: Making a New Science* (Viking, New York, USA).

Goldberger, A. L., Rigney, D. R., and West, B. J. (1990). Chaos and fractals in human physiology, *Sci. Am.*, 262(2), pp. 42–49.

Gopikrishnan, P., Plerou, V., Amaral, L. A. N., Meyer, M., and Stanley, H. E. (1999). Scaling of the distribution of fluctuations of financial market indices. http://xxx.lanl.gov/cond-mat/9905305.

Gribbin, J. (1985). *In Search of the Double Helix* (Wildwood House Ltd., London), p. 362.

Grosveld, F. and Fraser, P. (1997) Locus control of regions. In *Nuclear Organization, Chromatin Structure, and Gene Expression*, eds. Van Driel, R. and Otte, A. P. pp. 129–144 (Oxford University Press, Oxford).

Guharay, S., Hunt, B. R., Yorke, J. A., and White, O. R. (2000). Correlations in DNA sequences across the three domains of life, *Physica D*, 146, pp. 388–396. http://linkage.rockefeller.edu/wli/dna_corr/guharay00.pdf.

Gutenberg, B. and Richter, R. F. (1944). Frequency of earthquakes in California, *Bull. Seis. Soc. Amer.*, 34, p. 185.

Hao Bailin, Lee, H. and Zhang, S. (2000). Fractals related to long DNA sequences and complete genomes, *Chaos Solit. Fractals*, 11(6), pp. 825–836. http://linkage.rockefeller.edu/wli/dna_corr/haolee00.pdf.

Havlin S., Buldyrev S. V., Goldberger, A. L., Mantegna, R. N., Peng, C. K., Simons, M., and Stanley, H. E. (1995). Statistical and linguistic features of DNA sequences, *Fractals*, 3, pp. 269–264.

Herzel, H., Weiss, O., and Trifonov, E. N. (1998). Sequence periodicity in complete genomes of Archaea suggests positive supercoiling, *J. Biomol. Struct. Dynam.*, 16(2), pp. 341–345. http://linkage.rockefeller.edu/wli/dna_corr.

Herzel, H., Weiss, O., and Trifonov, E. N. (1999). 10-11 bp periodicities in complete genomes reflect protein structure and DNA folding, *Bioinformatics*, 15(3), pp. 187–193. http://linkage.rockefeller.edu/wli/dna_corr.

Hey, T. and Walters, P. (1989). *The Quantum Universe* (Cambridge University Press, Cambridge), p. 180.

Holste, D., Grosse, I., and Herzel, H. (2001). Statistical analysis of the DNA sequence of human chromosome 22, *Phys. Rev. E*, 64, pp. 041917(1–9). http://linkage.rockefeller.edu/wli/dna_corr/holste01.pdf.

Hooge, C., Lovejoy, S., Schertzer, D., Pecknold, S., Malouin, J. F., and Schmitt, F. (1994). Multifractal phase transitions: The origin of self-organized criticality in earthquakes, *Nonlinear Process. Geophysics*, 1, pp. 191–197.

Jenkinson, A. F. (1977). A Powerful Elementary Method of Spectral Analysis for use with Monthly, Seasonal or Annual Meteorological Time Series, Meteorological Office, London, Branch Memorandum No. 57, pp. 1–23.

Jean R. V. (1994). *Phyllotaxis: A Systemic Study in Plant Morphogenesis* (Cambridge University Press, New York).

Kadanoff, L. P. (1996). Turbulent excursions, *Nature*, 382, pp. 116–117.

Kane, R. P. (1996). Quasibiennial and quasitriennial oscillations in some atmospheric parameters, *PAGEOPH*, 147(3), pp. 567–583.

Kolmogorov, A. N. (1941). The local structure of turbulence in incompressible liquids for very high Reynolds numbers, *C. R. Russ. Acad. Sci.*, 30, pp. 301–305.

Kolmogorov, A. N. (1962). A refinement of previous hypotheses concerning the local structure of turbulence in a viscous inhomogeneous fluid at high Reynolds number, *J. Fluid Mech.*, 13, pp. 82–85.

Levine, D. and Steinhardt, J. (1984). Quasicrystals: A new class of ordered structures, *Phys.Rev.Letts.*, 53(26), pp. 2477–2480.

Leone, F. (1992) *Genetics: The Mystery and the Promise* (TAB Books, McGraw Hill, Inc.), p. 229.

Li, W. (1992). Generating nontrivial long-range correlations and $1/f$ spectra by replication and mutation, *Int. J. Bifur. Chaos*, 2(1), pp. 137–154. http://linkage.rockefeller.edu/wli/dna_corr/l-ijbc92-l.html.

Li, W. and Kaneko, K. (1992). Long-range correlation and partial $1/f^{\alpha}$ spectrum in a noncoding DNA sequence, *Europhys. Lett.*, 17(7), pp. 655–660. http://linkage.rockefeller.edu/wli/dna_corr/l-epl92-lk.html.

Li, W., Marr, T. G., and Kaneko, K. (1994). Understanding long-range correlations in DNA sequences, *Physica D*, 75(1–3), pp. 392–416; erratum: 82, p. 217 (1995). http://arxiv.org/chao-dyn/9403002.

Luderus, M. E. E. and van Driel, R. (1997). Nuclear matrix-associated regions. In *Nuclear Organization, Chromatin Structure, and Gene Expression*, eds. Van Driel, R. and Otte, A. P. pp. 99–115 (Oxford University Press, Oxford).

Maddox, J. (1988a). Licence to slang Copenhagen? *Nature*, 332, p. 581.

Maddox, J. (1988b). Turning phases into frequencies, *Nature*, 334, p. 99.

Maddox, J. (1993). Can quantum theory be understood? *Nature*, 361, p. 493.

Mandelbrot, B. B. (1975). On the geometry of homogenous turbulence with stress on the fractal dimension of the iso-surfaces of scalars, *J. Fluid Mech.*, 72, pp. 401–416.

Mandelbrot, B. B. (1977). *Fractals: Form, Chance and Dimension* (Freeman, San Francisco).

Mantegna, R. N. and Stanley, H. E. (1995). Scaling behaviour in the dynamics of an economic index, *Nature*, 376, pp. 46–49.

Mary Selvam, A. (1990). Deterministic chaos, fractals and quantumlike mechanics in atmospheric flows, *Can. J. Physics*, 68, pp. 831–841. http://xxx.lanl.gov/html/physics/0010046.

Mary Selvam, A., Pethkar, J. S., and Kulkarni, M. K. (1992). Signatures of a universal spectrum for atmospheric interannual variability in rainfall time series over the Indian Region, *Int'l J. Climatol.*, 12, pp. 137–152.

Mary Selvam, A. (1998). Quasicrystalline pattern formation in fluid substrates and phyllotaxis. In *Symmetry in Plants*, eds. Barabe, D. and Jean, R. V., Vol. 4 (World Scientific, Singapore), pp. 795–809. http://xxx.lanl.gov/abs/chao-dyn/9806001.

Mohanty, A. K. and Narayana Rao, A. V. S. S. (2000). Factorial moments analyses show a characteristic length scale in DNA sequences, *Phys. Rev. Lett.*, 84(8), pp. 1832–1835. http://linkage.rockefeller.edu/wli/dna_corr/mohanty00.pdf.

Monin, A. S. and Yaglom, A. M. (1975). *Statistical Hydrodynamics*. Vols. 1 and 2 (MIT Press, Cambridge).

Muller, A. and Beugholt, C. (1996). The medium is the message, *Nature*, 383, pp. 296–297.

Nagai, N., Kuwata, K., Hayashi, T., Kuwata, H., and Era, S. (2001). Evolution of the periodicity and the self-similarity in DNA sequence: A Fourier transform analysis, *Jap. J. Physiol.*, 51(2), pp. 159–168. http://linkage.rockefeller.edu/wli/dna_corr.

Nelson, D. R. (1986). Quasicrystals, *Sci. Amer.*, 255, pp. 42–51.

Newman, M. (2000). The power of design, *Nature*, 405, pp. 412–413.

Omori, F. (1895). On the aftershocks of earthquakes, *J. Coll. Sci.*, 7, p. 111.

Peng, C.-K., Buldyrev, S. V., Goldberger, A. L., Havlin, S., Sciortino, F., Simons, M., and Stanley, H. E. (1992). Long-range correlations in nucleotide sequences, *Nature*, 356, pp. 168–170. http://linkage.rockefeller.edu/wli/dna_corr/l-nature92-p.html.

Pizzi, E., Liuni, S., and Frontali, C. (1990). Detection of latent sequence periodicities, *Nucleic Acid. Res.*, 18(13), pp. 3745–3752. http://linkage.rockefeller.edu/wli/dna_corr.

Plerou, V., Gopikrishnan, P., Amaral, L. A. L., Meyer, M., and Stanley, H. E. (1999). Scaling of the distribution of price fluctuations of individual companies. http://xxx.lanl.gov/cond-mat/9907161.

Prabhu, V. V. and Claverie, J. M. (1992). Correlations in intronless DNA, Scientific Correspondence, *Nature*, 359, p. 782. http://linkage.rockefeller. edu/wli/dna_corr.

Rae, A. (1988). *Quantum-Physics: Illusion or Reality?* (Cambridge University Press, New York), p. 129.

Rae, A. I. M. (1999). Waves, particles and fullerenes, *Nature*, 401, pp. 651–653.

Richardson, L. F. (1960). The problem of contiguity: An appendix to statistics of deadly quarrels. In *General Systems — Year Book of the Society for General Systems Research*, eds. Von Bertalanffy, L. and Rapoport, A., Vol. V, pp. 139–187 (Ann Arbor, MI).

Richardson, L. F. (1965). *Weather Prediction by Numerical Process* (Dover, Mineola, New York).

Ruhla, C. (1992). *The Physics of Chance* (Oxford University Press, Oxford), p. 217.

Sambamurty, A. V. S. S. (1999). *Genetics* (Narosa Publishing House, New Delhi), p. 757.

Schrodinger, E. (1967). *What is Life?* (Cambridge University Press, Cambridge).

Schroeder, M. (1991). *Fractals,Chaos and Powerlaws* (W. H. Freeman and Co., New York).

Selvam, A. M. and Joshi, R. R. (1995). Universal spectrum for inter-annual variability in COADS global air and sea surface temperatures, *In'l. J. Climatol.*, 15, pp. 613–623.

Selvam, A. M., Pethkar, J. S., Kulkarni, M. K., and Vijayakumar, K. (1996). Signatures of a universal spectrum for atmospheric inter-annual variability in COADS surface pressure time series, *Int'l. J. Climatol.*, 16, pp. 393–404.

Selvam, A. M. and Fadnavis, S. (1998). Signatures of a universal spectrum for atmospheric inter-annual variability in some disparate climatic regimes, *Meteorol. Atmos. Phys.*, 66, pp. 87–112. http://xxx.lanl.gov/abs/chao-dyn/ 9805028.

Selvam, A. M. and Fadnavis, S. (1998). Cantorian fractal patterns, quantum-like chaos and prime numbers in atmospheric flows, *Chaos Solit Fractals*, http:// xxx.lanl.gov/abs/chao-dyn/9810011.

Selvam, A. M. (2001a). Quantum-like chaos in prime number distribution and in turbulent fluid flows. *APEIRON*, 8(3), pp. 29–64. http://xxx.lanl.gov/html/ physics/0005067, http://redshift.vif.com/JournalFiles/V08NO3PDF/ V08N3SEL.PDF.

Selvam, A. M. (2001b). Signatures of quantum-like chaos in spacing intervals of non-trivial Riemann zeta zeros and in turbulent fluid flows, *APEIRON*, 8(4), pp. 10–40. http://xxx.lanl.gov/html/physics/0102028, http://redshift.vif.com/ JournalFiles/V08NO4PDF/V08N4SEL.PDF.

Shiba, K., Takahashi, Y., and Noda, T. (2002). On the role of periodism in the origin of proteins, *J. Mol. Biol.*, 320(4), pp. 833–840.

Simon, R., Kimble, H. J., and Sudarshan, E. C. G. (1988). Evolving geometric phase and its dynamical interpretation as a frequency shift: An optical experiment, *Phys. Rev. Letts.*, 61(1), pp. 19–22.

Skinner, J. E. (1994). Low dimensional chaos in biological systems, *Bio. Technol.*, 12, pp. 596–600.

Som, A., Chattopadhyay, Chakrabarti, J., and Bandyopadhyay, D. (2001). Codon distributions in DNA, *Phys. Rev. E*, 63, pp. 1–8. http://linkage.rockefeller. edu/wli/dna_corr/som01.pdf.

Sornette, D., Johansen, A., and Bouchaud, J-P. (1995). Stock market crashes, precursors and replicas. http://xxx.lanl.gov/pdf/cond-mat/9510036.

Stanley, H. E. (1995). Power-laws and universality, *Nature*, 378, p. 554.

Stanley, M. H. R., Amaral, L. A. N., Buldyrev, S. V., Havlin, S., Leschhorn, H. Maass, P., Salinger, M. A., and Stanley, H. E. (1996a). Can statistical physics contribute to the science of economics? *Fractals*, 4(3), pp. 415–425.

Stanley, H. E., Amaral, L. A. N., Buldyrev, S. V., Goldberger, A. L., Havlin, S., Hyman, B. T., Leschhorn, H., Maass, P., Makse, H. A., Peng, C.-K., Salinger, M. A., Stanley, M. H. R., and Vishwanathan, G. M. (1996b). Scaling and universality in living systems, *Fractals*, 4(3), pp. 427–451.

Stanley, H. E., Afanasyev, V., Amaral, L. A. N., Buldyrev, S. V., Goldberger, A. L., Havlin, S., Leschhorn, H., Maass, P., Mantegna, R. N., Peng, C.-K., Prince, P. A., Salinger, M. A., Stanley, M. H. R., and Viswanathanan, G. M. (1996c). Anomalous fluctuations in the dynamics of complex systems: From DNA and physiology to econophysics, *Phys. A*, 224(1–2), pp. 302–321.

Stanley, H. E. (2000). Exotic statistical physics: Applications to biology, medicine, and economics, *Phys. A*, 285, pp. 1–17.

Stanley H. E., Amaral, L. A. N., Gopikrishnan, P., and Plerou, V. (2000). Scale invariance and universality of economic fluctuations, *Phys. A*, 283, pp. 31–41.

Stewart, I. (1992). Where do nature's patterns come from? *New Sci.*, 135, p. 14.

Stewart, I. (1995). Daisy, daisy, give your answer do, *Sci. Amer.*, 272, pp. 76–79.

Strathmann, R. R. (1990). Testing size abundance rules in a human exclusion experiment, *Science*, 250, p. 1091.

Thompson, D. W. (1963). *On Growth and Form*, 2nd Ed. (Cambridge University Press).

Voss, R. F. (1992). Evolution of long-range fractal correlations and 1/f noise in DNA base sequences, *Phys. Rev. Lett.*, 68(25), pp. 3805–3808.

Voss, R. F. (1994). Long-range fractal correlations in DNA introns and exons, *Fractals*, 2(1), pp. 1–6.

Wallace, J. M. and Hobbs, P. V. (1977) *Atmospheric Science: An Introductory Survey* (Academic Press, New York).

Watson, J. D. and Crick, F. H. C. (1953). A structure for deoxyribose nucleic acid, *Nature*, 25 April, pp. 737–738.

Watson, J. D. (1997). *The Double Helix* (Weidenfeld and Nicolson, London) p. 175.

West, B. J. (1990a). Fractal forms in physiology, *Int'l. J. Modern Physics B*, 4(10), pp. 1629–1669.

West, B. J. (1990b). Physiology in fractal dimensions, *Annal. Biomed. Eng.*, 18, pp. 135–149.

Widom, J. (1996). Short-range order in two eukaryotic genomes: Relation to chromosome structure, *J. Mol. Biol.*, 259, pp. 579–588. http://linkage.rockefeller.edu/wli/dna_corr.

Yu, Z-G., Anh, V. V., and Wang, B. (2000). Correlation property of length sequences based on global structure of the complete genome, *Phys. Rev. E*, 63, pp. 011903(1–8). http://linkage.rockefeller.edu/wli/dna_corr/yu00.pdf.

# Chapter 7

# Long-Range Correlations Data Set V: Universal Spectrum for DNA Base CG Frequency Distribution in *Takifugu Rubripes* (Puffer fish) Genome*

## 7.1 Introduction

The DNA bases A, C, G, T exhibit long-range spatial correlations manifested as inverse power law form for power spectra of spatial fluctuations
(Voss, 1992). Such non-local connections are intrinsic to the self-similar
fractal space-time fluctuations exhibited by dynamical systems in nature,
now identified as self-organized criticality, an intensive field of research
in the new discipline of nonlinear dynamics and chaos. The physics of the
observed self-organized criticality or the ubiquitous $1/f$ spectra found in
many disciplines of study is not yet identified (Milotti, 2002). A recently
developed general systems theory model (Selvam, 1990, 2007) predicts
the observed non-local connections as intrinsic to quantum-like chaos
governing the space-time evolution of dynamical systems in nature.
Identification of long-range correlations in the frequency distribution of
the bases A, C, G, T in the DNA molecule imply mutual communication

---

* http://arxiv.org/pdf/0704.2114 (2015). *Chaos and Complexity Letters*, 9(1), pp. 15–42.

and control between coding and non-coding DNA bases for cooperative organization of vital functions for the living system by the DNA molecule as a unified whole entity. In this chapter, it is shown that the power spectra of DNA bases CG concentration of *Takifugu rubripes* (Puffer fish) genome exhibits model predicted inverse power law form implying long-range correlations in the spatial distribution of the DNA bases CG concentration. *Takifugu rubripes* (Puffer fish) genome is about nine times smaller than the human genome with the same number of genes and therefore with less number of non-coding DNA and is therefore of special interest for identification of location of known and unknown genes in the human genome. The chapter is organized as follows: The applicability of the statistical normal distribution for analysis of fractal fluctuations is discussed in Section 7.1. Section 7.2 describes a multidisciplinary approach for modelling biological complexity and summarizes the general systems theory concepts and model predictions of universal properties characterizing the form and function of dynamical systems. The general systems theory model predicts quantum-like mechanical laws applicable to all dynamical systems. Section 7.3 gives details of the *Takifugu rubripes* (Puffer fish) DNA data sets and the analysis techniques used for the study. Section 7.4 contains the results and discussions of analysis of the data sets. Section 7.5 relates to current status of basic foundations of quantum mechanics with regard to real world phenomena. Section 7.6 states the important conclusions of the present study, in particular, the role of the apparently redundant non-coding DNA which comprises more than 90% of the human genome.

### 7.1.1  *Fractal fluctuations and statistical normal distribution*

Statistical and mathematical tools are used for analysis of data sets and estimation of the probabilities of occurrence of events of different magnitudes in all branches of science and other areas of human interest. Historically, the statistical normal or the Gaussian distribution has been in use for nearly 400 years and gives a good estimate for probability of occurrence of the more frequent moderate sized events of magnitudes within two standard deviations from the mean. The Gaussian distribution is based on the concept of data independence, fixed mean and standard deviation with a majority of data events clustering around the mean. However, for real world infrequent hazardous extreme events of magnitudes greater than two standard deviations, the statistical normal distribution gives progressively increasing under-estimates of up to near zero probability. In the 1890s the power law or Pareto distributions with implicit long-range correlations were found to fit the fat tails exhibited by hazardous extreme events such as heavy rainfall, stock market crashes, traffic jams, the after-shocks following major earthquakes, etc. A historical review of statistical normal and the Pareto distributions are given by Andriani and McKelvey (2007) and Selvam (2009). The spatial and/or temporal data sets in practice refer to real world or computed dynamical systems and are fractals with self-similar geometry and long-range correlations in space and/or time, i.e., the statistical properties such as the mean and variance are scale-dependent and do not possess fixed mean and variance and therefore the statistical normal distribution cannot be used to quantify/describe self-similar data sets. Though the observed power-law distributions exhibit qualitative universal shape, the exact physical mechanism underlying such scale-free power-laws is not yet identified for the formulation of universal quantitative equations for fractal fluctuations of all scales. The universal inverse power-law for fractal fluctuations is shown to be a function of the golden mean based on general systems theory concepts for fractal fluctuations (Chapter 2).

## 7.2 Multidisciplinary Approach for Modelling Biological Complexity

Computational biology involves extraction of the hidden patterns from huge quantities of experimental data, forming hypotheses as a result, and simulation based analyses, which tests hypotheses with experiments, providing predictions to be tested by further studies. Robust systems maintain their state and functions against external and internal perturbations, and robustness is an essential feature of biological systems. Structurally stable network configurations in biological systems increase insensitivity to parameter changes, noise and minor mutations (Kitano, 2002). Systems biology advocates a multidisciplinary approach for modelling biological complexity. Many features of biological complexity result from self-organization. Biological systems are, in general, global patterns produced by local interactions. One of the appealing aspects of the study of self-organized systems is that we do not need anything specific from biology to understand the existence of self-organization. Self-organization occurs for reasons that have to do with the organization of the interacting elements (Cole, 2002). The first and most general criterion for systems thinking is the shift from the parts to the whole. Living systems are integrated wholes whose properties cannot be reduced to those of smaller parts (Capra, 1996). Many disciplines may have helpful insights to offer or useful techniques to apply to a given problem, and to the extent that problem-focused research can bring together practitioners of different disciplines to work on shared problems, this can only be a good thing. A highly investigated but poorly understood phenomena, is the ubiquity of the so-called $1/f$ spectra in many interesting phenomena, including biological systems (Wooley and Lin, 2005).

## 7.2.1  *General systems theory for fractal fluctuations in dynamical systems*

The power (variance) spectra of fractal fluctuations exhibit inverse power law form $f^{-\alpha}$ indicating (i) the observed fluctuations result from the superimposition of an underlying eddy continuum fluctuation structure (ii) fractal fluctuations are self-similar, i.e., the larger and smaller scale fluctuation amplitudes are related to each other by the scale factor $\alpha$ alone independent of other details of the generating mechanism.

The general systems theory model is based on Townsend's (1956) concept that large eddy circulation can be visualized as envelope enclosing smaller scale fluctuations, i.e., large eddy is the integrated mean of enclosed smaller eddy circulations. The eddy continuum forms by space-time integration of successively larger scales. The eddy continuum growth occurs in two stages: (i) generation of dominant turbulent eddy of radius $r$ (ii) large eddies form as envelopes enclosing these dominant turbulent eddies starting from unit primary eddy as zero level with length scale ratio

$z$ ($R/r$) equal to 1, 2, 3, etc., for successive stages of eddy growth. The primary eddy growth region is $z = 0$ to $\pm 1$. The following quantitative characteristics of the eddy continuum are obtained.

(a)  The large eddy circulation speed $W$ of radius $R$ is expressed in terms of enclosed small eddy circulation speed $w_*$ of radius $r$ as

$$W^2 = \frac{2}{\pi} \frac{r}{R} w_*^2 \tag{7.1}$$

(b)  The eddy continuum traces an overall logarithmic spiral trajectory with the quasiperiodic *Penrose* tiling pattern (Fig. 7.1) for the internal structure. The successively larger eddy radii and the corresponding circulation speeds follow the *Fibonacci* number series. The ratio of successive radii lengths is equal to the golden mean $\tau$ ($\approx 1.618$). The relationship between the circulation speeds $W$ and $w_*$, respectively, of large and inherent small eddies of respective radii $R$ and $r$ is given by the logarithmic relationship

$$W = \frac{w_*}{k} \ln z \tag{7.2}$$

(c)  In Eq. (7.2), $k$ is the steady state fractional volume dilution of the large eddy by internal smaller scale eddy mixing (Selvam, 2013, 2014a) and is given as

$$k = \sqrt{\frac{\pi}{2z}} \tag{7.3}$$

(d)  By concept the large eddy circulation speed $W$ is the integrated mean of enclosed small-scale circulation speeds $w_*$. Therefore at each level, $W$ represents the mean value associated with standard deviation equal to $w_*$. The normalized deviation $t$ equal to mean/standard deviation is now given by $W/w_*$ which is proportional to $\ln z$ from the aforementioned logarithmic relationship (Eq. (7.2)). For fixed small eddy radius $r$, the normalized deviation $t$ represents logarithm of eddy wavelengths (frequencies) and is given as

$$t = \frac{\log L}{\log T_{50}} - 1 \qquad (7.4)$$

In Eq. (7.4) $L$ is the time period (or wavelength) and $T_{50}$ is the period up to which the cumulative percentage contribution to total variance is equal to 50 and $t = 0$. $\log T_{50}$ also represents the mean value for the r.m.s. eddy fluctuations and is consistent with the concept of the mean level represented by r.m.s. eddy fluctuations. The probability $P$ distributions of amplitudes and/or variance of eddy fluctuations when plotted with normalized deviation $t$ will represent the conventional plot with logarithm of eddy wavelength (frequency) on $x$-axis versus probability $P$ on $y$-axis. Such a plot of probability distributions $P$ of amplitude and also variance of fractal fluctuations displays inverse power-law form discussed at items (e) and (f) in the following.

(e)  The probability distribution $P$ (of amplitude as well as variance of fractal fluctuations) is a function of the golden mean $\tau$. The golden mean which underlies the *Fibonacci* series represents a nested continuum of self-similar structures or fractals implying long-range correlations or persistence between the larger and smaller scale structures and is consistent with observed inverse power law distribution for $P$. For the range of normalized deviation $t$ values $t \geq 1$ and $t \leq -1$, $P$ is given as

$$P = \tau^{-4t} \qquad (7.5)$$

(f)  The primary eddy growth corresponds to normalized deviation $t$ ranging from $-1$ to $+1$. In this region the probability $P$ is a function of the steady state fractional volume dilution $k$ of the growing primary eddy by internal smaller scale eddy mixing (Selvam, 2013, 2014a) (Eq. (7.3)). The expressions for $P$ is given as

$$P = \tau^{-4k} \qquad (7.6)$$

(g)  The model predicted universal inverse power-law distribution is very close to statistical normal distribution for normalized deviation $t < 2$ and $t > -2$ and exhibits a long fat tail for $t \geq 2$ and $t \leq 2$, i.e., extreme events have a higher probability of occurrence than that predicted by statistical normal distribution as found in practice (Selvam, 2013, 2014a).

(h) The general systems theory model shows that (i) the probability distribution *P* as given in items (d)–(f) when plotted with respect to normalized deviation *t* gives unique and universal quantification for the observed inverse power-law distribution for power (variance) spectrum of fractal fluctuations (ii) the amplitude distribution of fractal fluctuations is also quantified by the same probability distribution *P* (Selvam, 2013, 2014a). Therefore fractal fluctuations signify quantum-like chaos since the property that the additive amplitude of eddies when squared represent the probability densities is exhibited by the subatomic dynamics of quantum systems such as the electron or photon (a brief review of current status quantum mechanics is given in Section 7.5).

The general systems theory concepts are equivalent to *Boltzmann's* postulates and the *Boltzmann distribution* (for molecular energies) with the inverse power law expressed as a function of the golden mean is the universal probability distribution function for (the amplitude as well as variance) of observed fractal fluctuations which corresponds closely to statistical normal distribution for moderate amplitude fluctuations and exhibit a fat long tail for hazardous extreme events in dynamical systems (Selvam, 2014a). The general systems theory model predictions given previously are based on classical statistical physical concepts and satisfies the principle of maximum entropy production for dynamical system in steady-state equilibrium (Selvam, 2013).

### 7.2.2 Fractals represent hierarchical communication networks

The evolution of dynamical systems is governed by ordered information communication between the small-scale internal networks and the overall large scale growth pattern. The hierarchical network architecture underlying fractal space-time fluctuations (Fig. 7.1) provides robust two-way information communication and control for integrity in the performance of vital functions specific to the dynamical system.

Complex networks from such different fields as biology, technology or sociology share similar organization principles. The possibility of a unique growth mechanism promises to uncover universal origins of collective behaviour. In particular, the emergence of self-similarity in complex networks raises the fundamental question of the growth process according to which these structures evolve (Song *et al.*, 2006).

### 7.2.3  *Model predictions (relevance of model predictions to biological networks)*

The general systems theory model predictions for the space-time fractal fluctuation pattern of dynamical systems (Selvam, 1990, 2007) are given in the following.

#### 7.2.3.1  *Quasicrystalline pattern for the network architecture*

The apparently irregular fractal fluctuations can be resolved into a precise geometrical pattern with logarithmic spiral trajectory and the quasi periodic Penrose tiling pattern (Steinhardt, 1997; Baake, 2002) for the internal structure (Fig. 7.1) on all scales to form a nested continuum of vortex roll circulations with ordered energy flow between the scales (Selvam, 2007). Logarithmic spiral formation with Fibonacci winding number and five-fold symmetry possess maximum packing efficiency for component parts and are manifested strikingly in plant *Phyllotaxis* (Jean, 1994; Selvam, 1998).

Aperiodic or quasiperiodic order is found in different domains of science and technology. There is widespread presence of *Fibonacci* numbers and the golden mean in different physical contexts (Boeyens and Thackeray, 2014). Several conceptual links exist between quasiperiodic crystals and the hierarchical structure of biopolymers in connection with the charge transfer properties of both biological and synthetic DNA chains. DNA was originally viewed as a trivially periodic macromolecule, unable to store the amount of information required for the governance of cell function. The famous physicist, Schrodinger (1945), was the first to

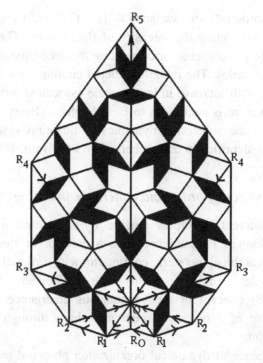

**Figure 7.1:**    Internal structure of large eddy circulations.

suggest that genetic material should consist of a long sequence of a few repeating elements exhibiting a well-defined order without the recourse of periodic repetition, thereby introducing the notion of aperiodic crystal. Nevertheless, this notion remained dormant for almost four decades until the discovery of quasicrystalline solids. An essential characteristic of the quasicrystalline order is self-similarity which reveals the existence of certain motifs in the sample which contain the whole structure enfolded within them (Macia, 2006).

### 7.2.3.2 *Long-range spatiotemporal correlations (coherence)*

The overall logarithmic spiral pattern enclosing the internal small-scale closed networks $OR_0R_1$, $OR_1R_2$, ... may be visualized as a continuous smooth rotation of the phase angle $\theta$ ($R_0OR_1$, $R_0OR_2$, ... etc.) with increase in period. The phase angle $\theta$ for each stage of growth is equal to

$r/R$ and is proportional to the variance $W^2$ (Eq. (7.1), Selvam, 1990, 2007), the variance representing the intensity of fluctuations. The phase angle gives a measure of coherence or correlation in space-time fluctuations of different length scales. The model predicted continuous smooth rotation of phase angle with increase in length scale associated with logarithmic spiral flow structure is analogous to Berry's phase (Berry, 1988; Kepler *et al.*, 1991) in the subatomic dynamics of quantum systems. Berry's phase has been identified in atmospheric flows (Selvam, 1990, 2007).

### 7.2.3.3 *Emergence of order and coherence in biology*

Macroscale coherent structures emerge by space-time integration of microscopic domain fluctuations. Such a concept of the autonomous growth of hierarchical network continuum with ordered energy flow between the scales is analogous to Prigogine and Stenger's (Prigogine and Stengers, 1988) concept of the spontaneous emergence of order and organization out of apparent disorder and chaos through a process of self-organization.

The emergence of dynamical organization observed in physical and chemical systems should be of importance to biology (Karsenti, 2008). Underlying the apparent complexity (of living matter), there are fundamental organizational principles based on physicochemical laws (Karsenti, 2007).

The general systems theory for coherent pattern formation summarized earlier may provide a model for biological complexity. General systems theory is a logical-mathematical field, the subject matter of which is the formulation and deduction of those principles which are valid for "systems" in general, whatever the nature of their component elements or the relations or "forces" between them (Von Bertalanffy, 1968; Klir, 1992; Peacocke, 1989).

### 7.2.3.4 *Dominant length scales in the quasicrystalline spatial pattern*

The eddy continuum underlying fractal fluctuations has embedded robust dominant wavebands $R_0OR_1$, $R_1R_2O$, $R_3R_2O$, $R_3R_4O$, ...... with length (time) scales $T_D$ which are functions of the golden mean $\tau$ and the primary eddy energy perturbation length (time) scale $T_S$ such as the annual cycle of summer to winter solar heating in atmospheric flows. The dominant eddy length (time) scale for the $n$th dominant eddy is given as

$$T_D = T_S(2 + \tau)\tau^n \qquad (7.7)$$

The successive dominant eddy length (time) scales for unit primary perturbation length (time) scale, i.e., $T_S = 1$, are given (Selvam, 1990, 2007) as 2.2, 3.6, 5.9, 9.5, 15.3, 24.8, 40.1, 64.9, ... respectively for values of $n = -1, 0, 1, 2, 3, 4, 5, 6,...$

Space-time integration of small-scale internal circulations results in robust broadband dominant length scales which are functions of the primary length scale $T_S$ alone and are independent of exact details (chemical, electrical, physical, etc.) of the small-scale fluctuations. Such global scale spatial oscillations in the unified network are not affected appreciably by failure of localized microscale circulation networks (Kitano, 2002).

Wavelengths (or periodicities) close to the model predicted values have been reported in weather and climate variability, prime number distribution (Selvam, 2014b), Riemann zeta zeros (non-trivial) distribution, Drosophila DNA base sequence, stock market economics, Human chromosome 1 DNA base sequence (Selvam, 2007 and all references therein).

### 7.2.3.5 DNA sequence and functions

The double-stranded DNA molecule is held together by chemical components called bases; Adenine (A) bonds with thymine (T); cytosine (C) bonds with guanine (G) These letters form the "code of life"; there are close to three billion base pairs in mammals such as humans and rodents. Written in the DNA of these animals are 25,000–30,000 genes which cells use as templates to start the production of proteins; these sophisticated molecules build and maintain the body. According to the traditional viewpoint, the really crucial things were genes, which code for proteins — the "building blocks of life". A few other sections that regulate gene function were also considered useful. The rest was thought to be excess baggage — or "junk" DNA. But new findings suggest this interpretation may be wrong. Comparison of genome sequences of man, mouse and rat and also chicken, dog and fish sequences show that several great stretches of DNA were identical across the species which shared an ancestor about 400 million years ago. These "ultra-conserved" regions do not appear to code for protein, but obviously are of great importance for survival of the animal. Nearly a quarter of the sequences overlaps with genes and may help in protein production. The conserved elements that do not actually overlap with genes tend to cluster next to genes that play a role in embryonic development (Kettlewell, 2004; Check, 2006).

A number of the elements in the non-coding portion of the genome may be central to the evolution and development of multicellular organisms. Understanding the functions of these non-protein-coding sequences, not just the functions of the proteins themselves, will be vital to understanding the genetics, biology and evolution of complex organisms (Taft and Mattick, 2003).

Brenner *et al.* (1993) proposed as a model for genomic studies the tiger pufferfish (*Taki*) *Fugu rubripes*, a marine pufferfish with a genome nine times more compact than that of human. *Fugu rubripes* is separated from *Homo sapiens* by about 450 million years of evolution. Many comparisons have been made between *F. rubripes* and human DNA that demonstrate the potential of comparative genomics using the pufferfish genome (Crollius *et al.*, 2000). *Fugu*'s genome is compact, with less than 20% made up of repetitive sequences and fully one-third occupied by gene loci. Aparicio *et al.* (2002) report the sequencing and initial analysis of the *Fugu* genome. The usefulness of vertebrate comparative genomics was demonstrated by the identification of about thousand new genes in the human genome through comparison with *Fugu*. Functional comparisons between different fish genomes and the genomes of higher vertebrates will shed new light on vertebrate evolution and lead to the identification of the genes that distinguish fish and humans.

In the present study, it is shown that long-range spatial correlations are exhibited by *Fugu rubripes* DNA base C+G concentration per 10 bp (base pair) along different lengths of DNA containing both coding and non-coding sequences indicating global sequence control over coding functions.

The study of C+G variability in genomic DNA is of special interest because of the documented association of gene dense regions in GC rich DNA sequences. GC-rich regions include many genes with short introns while GC-poor regions are essentially deserts of genes. This suggests that the distribution of GC content in mammals could have some functional relevance, raising the issue of its origin and evolution (Galtier *et al.*, 2001; Semon *et al.*, 2005).

## 7.3 Data and Analysis

### 7.3.1 *Data*

The draft sequence of *Takifugu rubripes* (Puffer fish) genome assembly release 4 was obtained from "The Fugu Informatics Network" (ftp://fugu. biology.qmul.ac.uk/pub/fugu/scaffolds_4.zip) at School of Biological & Chemical Sciences, Queen Mary, University of London. The fourth assembly of the Fugu genome consists of 7,000 scaffolds. The individual contig size ranges from 2 to 1,100 Kbp (kilo base pairs), nearly half the genome in just 100 scaffolds, 80% of the genome in 300 scaffolds.

Non-overlapping DNA sequence lengths without breaks (N values) were chosen and then grouped in two categories A and B; category A consists of DNA lengths greater than 3 Kbp but less than 30 Kbp and category B consists of DNA sequence lengths greater than 30 Kbp. The data series were subjected to continuous periodogram power spectral analyses. For convenience in averaging the results, the Category A data lengths were grouped into seven groups and category B data into 18 groups such that the total number of bases in each group is equal to about 12 Mbp (million base pairs).

### 7.3.2 *Power spectral analyses: Variance and phase spectra*

The number of times base C and also base G, i.e., (C+G), occur in successive blocks of 10 bases were determined in the DNA length sections giving a C+G frequency distribution series of 300–3,000 values in category

A and more than 3,000 for category B. The power spectra of frequency distribution of C+G bases (per·10 bp) in the data sets were computed accurately by an elementary, but very powerful method of analysis developed by Jenkinson (1977) which provides a quasicontinuous form of the classical periodogram allowing systematic allocation of the total variance and degrees of freedom of the data series to logarithmically spaced elements of the frequency range (0.5, 0). The cumulative percentage contribution to total variance was computed starting from the high frequency side of the spectrum. The power spectra were plotted (Section 7.2.1, Eq. (7.4)) as cumulative percentage contribution to total variance versus the normalized standard deviation $t$ equal to $(\log L/\log T_{50})-1$

where $L$ is the period in years and $T_{50}$ is the period up to which the cumulative percentage contribution to total variance is equal to 50 (Selvam, 1990, 2007). The corresponding phase spectra were computed as the cumulative percentage contribution to total rotation (Section 7.2.3.1). The statistical chi-square test (Spiegel, 1961) was applied to determine the "goodness of fit" of variance and phase spectra with statistical normal as well as model predicted distributions. The variance and phase spectra were considered to be the same as model predicted/normal spectrum for regions ($t$-values) where the model predicted/normal spectrum lies within two standard deviations from mean variance and phase spectrum.

## 7.4 Results and Discussion

### 7.4.1 *Data sets and power spectral analyses*

The average, standard deviation, maximum, minimum and median DNA lengths (bp) for each group in the two data categories A and B are shown in Fig. 7.2. It is seen that the mean is close to the median and almost constant for the different data groups, particularly for category B (DNA length > 30 Kbp).

Figure 7.3 gives for each group in the two categories A and B, (i) the average and standard deviation of CG density per 10 bp (ii) the average and standard deviation of $T_{50}$ in length unit 10bp, (iii) the percentage number of variance spectra (V) and phase spectra (P) following normal distribution and the number of data sets.

## Spectral analyses - DNA base CG frequency distribution
### Takifugu rubripes (Puffer fish) Release 4

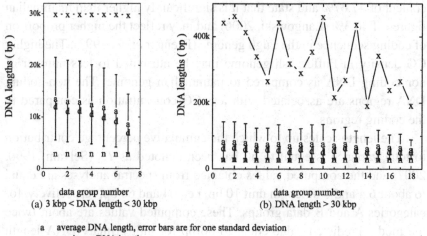

(a) 3 kbp < DNA length < 30 kbp

(b) DNA length > 30 kbp

a   average DNA length, error bars are for one standard deviation
x   maximum DNA length
n   minimum DNA length
d   median DNA length

**Figure 7.2:** Details of data sets used for analyses.

## Spectral analyses DNA base CG frequency distribution
### Takifugu rubripes (Puffer fish) Release 4 - Details of data sets and averaged results

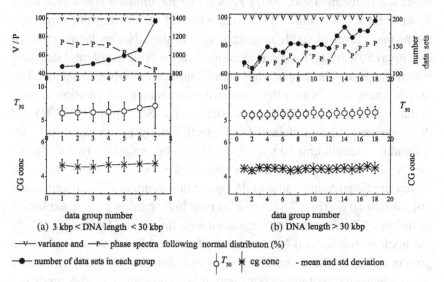

(a) 3 kbp < DNA length < 30 kbp

(b) DNA length > 30 kbp

—v— variance and —P— phase spectra following normal distributon (%)

—●— number of data sets in each group      $T_{50}$   ✳ cg conc   - mean and std deviation

**Figure 7.3:** Details of data sets and averaged results of spectral analyses.

The average CG density per 10 bp is about 47% and 45%, respectively, for categories A and B data groups. Elgar *et al.* (1999) report a G+C content of 47.67% and state that it is significantly higher than mammalian figures of 40.3% (Langowski, 2006) and may reflect the higher proportion of coding sequence in the Fugu genome (Brenner *et al.*, 1993). The higher CG density in Puffer fish genome may be attributed to less number of non-coding DNA as compared to mammalian genome. The non-coding DNA regions are associated with less CG concentration as compared to the coding regions.

The length scales up to which the cumulative percentage contribution to total variance is equal to 50 has been denoted as $T_{50}$ (Selvam, 1990, 2007) and the computed values obtained from spectral analyses are equal to about 6.4 and 6 in length unit 10 bp, i.e., 64 and 60 bp, respectively, for categories A and B data groups. These computed values are about twice the model predicted value equal to 36 bp for unit primary DNA length scale $OR_O = T_S = 10$ bp (Fig. 7.1). The observed higher value may be explained as follows. The first level of organized structure in the DNA molecule is the nucleosome formed by two turns of about 146 bp around a histone protein octamer (Bates and Maxwell, 1993; Skipper, 2006). Each turn of the nucleosome contains 73 bp and may represent a model predicted (Selvam, 1990, 2007) $T_{50} \approx 60$ bp for fundamental length scale $T_S = 10\tau$ bp $= OR_1$, (Fig. 7.1). A single turn of the nucleosome structure, the primary level of stable organized unit of the DNA molecule contributes up to 50% of the total variance signifying its role as internal structure which makes up the larger scale architecture of the DNA molecule. The length scales up to which the cumulative percentage contribution to total variance is equal to 50 has been denoted as $T_{50}$ (Selvam, 1990, 2007) and the computed values obtained from spectral analyses are equal to about 6.4 and 6 in length unit 10 bp, i.e., 64 and 60 bp, respectively, for categories A and B data groups. These computed values are about twice the model predicted value equal to 36 bp for unit primary DNA length scale $OR_O = T_S = 10$ bp (Fig. 7.1). The observed higher value may be explained as follows. The first level of organized structure in the DNA molecule is the nucleosome formed by two turns of about 146 bp around a histone protein octamer (Bates and Maxwell, 1993; Skipper, 2006). Each turn of the nucleosome contains 73 bp and may represent a model predicted

(Selvam, 1990, 2007) $T_{50} \approx 60$ bp for fundamental length scale $T_S = 10\tau$ bp = $OR_1$ (Fig. 7.1). A single turn of the nucleosome structure, the primary level of stable organized unit of the DNA molecule contributes up to 50% of the total variance signifying its role as internal structure which makes up the larger-scale architecture of the DNA molecule.

The variance spectra follow the model predicted and the statistical normal distribution in almost all the data groups in both the data categories, the model predicted spectrum being close to statistical normal distribution for normalized deviation $t$ values less than 2 on either side of $t = 0$ (Section 7.2.1, item g). Li and Holste (2005) have identified universal $1/f$ noise, crossovers of scaling exponents, and chromosome-specific patterns of guanine-cytosine content in DNA sequences of the human genome. A total average of about 63.7% and 72% of the data sets (Fig. 7.3), respectively, in categories A and B show that phase spectra follow the statistical normal distribution.

The average variance and phase spectra along with standard deviations for the data groups in the two categories A and B and the statistical normal distribution are shown in Fig. 7.4. The average variance spectra follow closely the statistical normal distribution for categories A and B data groups. The average phase spectra for category B data groups alone follow closely the statistical normal distribution while average phase spectra for category A data groups show appreciable departure from statistical normal distribution in agreement with the percentage number of variance spectra (V) and phase spectra (P) following the universal inverse power law form of the statistical normal distribution shown in Fig. 7.3.

Poland (2004, 2005) reported that C–G distribution in genomes is very broad, varying as a power law of the size of the block of genome considered and they also find the power law form for the C–G distribution for all of the species treated and hence this behavior seems to be ubiquitous.

## 7.4.2 *Model predicted dominant wavebands*

The general systems theory predicts that the broadband power spectrum of fractal fluctuations will have embedded dominant wavebands, the bandwidth increasing with wavelength, and the wavelengths are functions

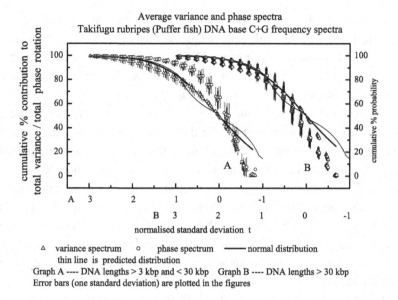

**Figure 7.4.**    The average variance and phase spectra of frequency distribution of bases C+G in *Takifugu rubripes* (Puffer fish) for the data sets given in Figs. 7.2 and 7.3. The power spectra were computed as cumulative percentage contribution to total variance versus the normalized standard deviation $t$ equal to $(\log L / \log T_{50}) - 1$ where $L$ is the wavelength in units of 10 bp and $T_{50}$ is the wavelength up to which the cumulative percentage contribution to total variance is equal to 50. The corresponding phase spectra were computed as the cumulative percentage contribution to total rotation (Section 7.2.3.1).

of the golden mean (Selvam, 1990, 2007). The first 13 values of the model predicted (Selvam, 1990, 2007) dominant peak wavelengths are 2.2, 3.6, 5.8, 9.5, 15.3, 24.8, 40.1, 64.9, 105.0, 167.0, 275, 445.0 and 720 in units of the block length 10 bp (base pairs) in the present study. The dominant peak wavelengths in Category A data sets with shorter DNA lengths (< 30 Kbp) were grouped into 17 class intervals 2–2.1, 2.1–2.3, 2.3–2.5, 2.5–3, 3–4, 4–6, 6–12, 12–20, 20–30, 30–50, 50–80, 80–120, 120–200, 200–300, 300–600, 600–1000 (in units of 10 bp block lengths) to include the model predicted dominant peak length scales mentioned earlier. For Category B data sets with longer DNA lengths (> 30 Kbp), the dominant peak wavelengths were grouped into 13 class intervals 2–3, 3–4, 4–6, 6–12, 12–20, 20–30, 30–50, 50–80, 80–120, 120–200, 200–300, 300–600, 600–1,000 (in units of 10 bp block lengths) to include the model predicted dominant peak length scales. Category A with shorter DNA lengths will exhibit more number of dominant wavebands in the shorter wavelengths and therefore more number of class intervals in the shorter wavelength region 2–3 (in units of 10 bp block lengths) for Category A. The class intervals increase in size progressively to accommodate model predicted increase in bandwidth associated with increasing wavelength.

Average class interval-wise percentage frequencies of occurrence of dominant wavelengths (normalized variance greater than 1) are shown in Fig. 7.5 along with the percentage contribution to total variance in each class interval corresponding to the normalized standard deviation $t$ (Selvam, 1990, 2007) computed from the average $T_{50}$ (Fig. 7.3). In this context it may be mentioned that statistical normal probability density distribution represents the eddy variance for values of $t$ less than 2 (Section 7.2.1 item g). The observed frequency distribution of dominant

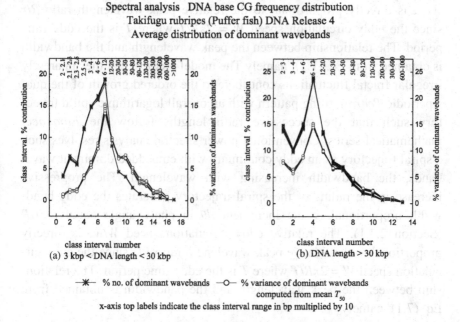

Spectral analysis  DNA base CG frequency distribution
Takifugu rubripes (Puffer fish) DNA Release 4
Average distribution of dominant wavebands

(a) 3 kbp < DNA length < 30 kbp

(b) DNA length > 30 kbp

—✳— % no. of dominant wavebands  —O— % variance of dominant wavebands computed from mean $T_{50}$
x-axis top labels indicate the class interval range in bp multiplied by 10

**Figure 7.5:** Dominant wavelengths in DNA bases C+G concentration distribution. Average class interval-wise percentage frequency distribution of dominant (normalized variance greater than 1) wavelengths is given by – * – (line + star). The corresponding computed percentage contribution to the total variance for each class interval is given by -O- (line + open circle). The observed frequency distribution of dominant eddies closely follow the model predicted computed percentage contribution to total variance

eddies follow closely the computed percentage contribution to total variance.

### 7.4.3 *Peak wavelength versus bandwidth*

The model predicts that the apparently irregular fractal fluctuations contribute to the ordered growth of the quasiperiodic Penrose tiling pattern with an overall logarithmic spiral trajectory such that the successive radii lengths follow the *Fibonacci* mathematical series. Conventional power spectral analysis resolves such a spiral trajectory as an eddy continuum with embedded dominant wavebands, the bandwidth increasing with wavelength. The progressive increase in the radius of the spiral trajectory generates the eddy bandwidth proportional to the increment $d\theta$ in phase angle equal to $r/R$ (Section 7.2.3.1). The relative eddy circulation speed $W/w_*$ is directly proportional to the relative peak wavelength ratio $R/r$ since the eddy circulation speed $W = 2\pi R/T$ where $T$ is the eddy time period. The relationship between the peak wavelength and the bandwidth is obtained from Eq. (7.1), namely The model predicts that the apparently irregular fractal fluctuations contribute to the ordered growth of the quasiperiodic *Penrose* tiling pattern with an overall logarithmic spiral trajectory such that the successive radii lengths follow the *Fibonacci* mathematical series. Conventional power spectral analysis resolves such a spiral trajectory as an eddy continuum with embedded dominant wavebands, the bandwidth increasing with wavelength. The progressive increase in the radius of the spiral trajectory generates the eddy bandwidth proportional to the increment $d\theta$ in phase angle equal to $r/R$ (Section 2.3.1). The relative eddy circulation speed $W/w_*$ is directly proportional to the relative peak wavelength ratio $R/r$ since the eddy circulation speed $W = 2\pi R/T$ where $T$ is the eddy time period. The relationship between the peak wavelength and the bandwidth is obtained from Eq. (7.1), namely

$$W^2 = \frac{2}{\pi}\frac{r}{R}w_*^2$$

Considering eddy growth with overall logarithmic spiral trajectory relative eddy bandwidth $\propto r/R$

The eddy circulation speed is related to eddy radius as

$$W = \frac{2\pi R}{T}$$

$W \propto R \propto$ peak wavelength

The relative peak wavelength is given in terms of eddy circulation speed as the relative peak wavelength is given in terms of eddy circulation speed as

relative peak wavelength $\propto \dfrac{W}{w_*}$

From Eq. (7.1) the relationship between eddy bandwidth and peak wavelength is obtained as

eddy bandwidth = (peak wavelength)$^2$

$$\frac{\log(eddy\ bandwidth)}{\log(peak\ wavelength)} = 2$$

A log–log plot of peak wavelength versus bandwidth will be a straight line with a slope (bandwidth/peak wavelength) equal to 2. A log–log plot of the average values of bandwidth versus peak wavelengths shown in Figs. 7.6 and 7.7 exhibits average slopes approximately equal to 2.5.

## 7.4.4 *Quasiperiodic Penrose tiling and packing efficiency*

Ten base pairs occur per turn of the DNA helix. Therefore, the fundamental length scale in the case of the DNA molecule is 10 base pairs forming a near complete circle (360 degrees) with 10-fold symmetry. Also, five-fold symmetry is exhibited by the carbon atoms of the sugar-phosphate backbone structure which supports the four bases (A, C, G, T) (Kettlewell, 2004). The 10-fold and 5-fold symmetries underlying the DNA architecture may signify spatial arrangement of the DNA bases in the mathematically precise ordered form of the quasiperiodic *Penrose* tiling pattern (Fig. 7.1).

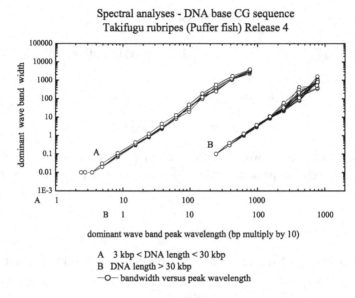

**Figure 7.6:** Log–log plot of peak wavelength versus bandwidth for dominant wavebands.

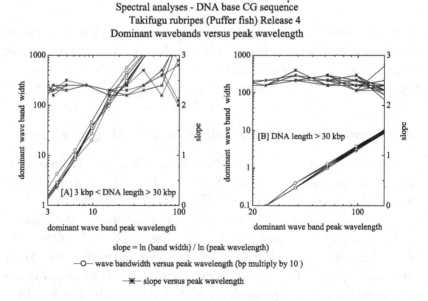

**Figure 7.7:** Same as in Fig. 7.6 for a limited peak wavelength range along with the corresponding computed slope equal to ln(bandwidth)/ln(peak wavelength).

Model predicted universal inverse power law followed by power spectra of CG fluctuations imply spontaneous organization of the DNA base sequence in the form of the quasicrystalline structure of the quasiperiodic Penrose tiling pattern (Fig. 7.1), characterized by a nested loop (coiled coil) structure. The Fibonacci sequence underlying the quasiperiodic Penrose tiling pattern is found in nature in plant *phyllotaxis* (Jean, 1994; Schrodinger, 1945). The model predicted quasicrystalline structure for the linear DNA string is associated with maximum packing efficiency (Selvam, 1998) and may help identify the geometry of the compact DNA structure inside the cell nucleus.

The dynamical architecture of the cell nucleus can be regarded as one of the "grand challenges" of modern molecular and structural biology. The genomic DNA and the histone proteins compacting it into chromatin comprise most of the contents of the nucleus. In every human cell, for instance, $6 \times 10^9$ base pairs of DNA, that is, a total length of about 2 m must be packed into a more or less spheroid nuclear volume about 10–20 $\mu$m in diameter (Langowski, 2006). The reported packing efficiency of the DNA string is then equal to about $10^5$ and is shown in the following to result from 10 stages of successive coiling.

The packing efficiency of the quasicrystalline structure is computed as follows. A length $L$ of DNA sequence in the approximately circular coiled form has a diameter $d$ equal to $L/p$ and therefore has a packing efficiency $P_{eff}$ equal to $d/L = \pi$. For $N$ stages of successive coiling, $P_{eff}$ is equal to $\pi N$ and for $N = 10$, $P_{eff} = 10^{4.97} \approx 10^5$, i.e., 10 stages of successive coiling results in five-fold compaction of the long DNA string inside the cell nucleus in agreement with reported values (Langowski, 2006).

# 7.5 Current Status of Basic Concepts in Quantum Mechanics

Quantum theory is based on a clear mathematical apparatus, has enormous significance for the natural sciences, enjoys phenomenal predictive success, and plays a critical role in modern technological developments. Yet, nearly 90 years after the theory's development, there is still no consensus in the scientific community regarding the interpretation of the theory's foundational building blocks (Schlosshauer *et al.*, 2013a, 2013b).

Classical mechanics (Newtonian mechanics) can explain macroscopic phenomena while quantum mechanics is used to explain microscopic dynamics of quantum systems such as the electron or photon. In Newtonian mechanics objects are solid particles and the laws are written in terms of precisely defined particle trajectories. Quantum mechanics is a collection of postulates based on a large number of observations. The following observed characteristics of microscopic scale quantum systems such as the electron or photon do not have satisfactory real world physical explanations (i) Quantum particles can act as both waves (eddies) and particles and possess wave-particle duality and as a result the position and momentum of the particle are indeterminate, i.e., can be given in terms of probabilities only. (ii) The square of wave amplitude gives the probability of occurrence of the quantum system at that location. (iii) Any measurement on quantum system disturbs the state of the system in an unpredictable manner. (iv) Energy of quantum particles is quantized. (v) The separated parts of a quantum system responds as a unified whole to local perturbations manifested as non-local connection.

In the following (Sections 7.5.1–7.5.3) it is shown that the general systems theory model for fractal fluctuations in dynamical systems (Selvam, 2007) predicts that the macroscale real world phenomena of weather systems exhibit the aforesaid listed quantum-like behavior as a natural consequence of intrinsic eddy (wave) continuum structure of atmospheric flows which functions as a unified whole communicating network with long-range space-time correlations.

Palmer (2005, 2009, 2014) has investigated the role of fractals in quantum theory (Section 7.1). The string-like energy flow pattern of atmospheric eddy continuum is similar to string theory for quantum phenomena put forth by Hooft (2014) (Section 7.2). Bush (2014) (Section 7.3) suggests fluid mechanical concepts may help explain quantum mechanical phenomena.

### 7.5.1 *Fractals and quantum theory*

The role of fractals in fundamental physics was investigated by Palmer (2005, 2009, 2014). Palmer (2005) discusses possible linkages between nonlinear paradigms developed to understand meteorological predictability and the deepest conceptual problems of quantum theory. Application of nonlinear meteorological thinking may indeed provide fresh insights on the foundational problems of quantum theory. Quantum theory is the most successfully tested, yet least well understood, of all physical theories (see, e.g., Penrose, 1989). Einstein's dissatisfaction with quantum theory is well known; the two key reasons for such dissatisfaction: indeterminacy and non-local causality. The notion that fractals may play a role in fundamental physics is not itself new. However, while earlier studies have focused on the concept that space-time itself may be fractal (Ord, 1983; Nottale and Schneider, 1984; El Naschie, 2004), Palmer considers the ontological significance of fractal geometry in state space.

In the 1960s, the introduction of global space-time geometric and topological methods, transformed our understanding of classical gravitational physics (Penrose, 1965). It is proposed that the introduction of global geometric and topological methods in state space may similarly transform our understanding of quantum gravitational physics. Combining these rather different forms of geometry may provide the missing element needed to advance the search for a unified theory of physics (Palmer, 2009). Arguments over quantum theory have raged since the 1920s. The mathematics of fractals may help to understand the long-standing puzzles of quantum theory. Very few scientists working on fundamental physics have explored how fractals might be incorporated into the theory, even though they are commonplace in other parts of physics (Buchanan, 2009). Palmer (2014) states that there were three great revolutions in 20th century theoretical physics: relativity theory, quantum theory and chaos theory. It is proposed that the three great revolutions of 20th century physics can be unified if the universe is considered a deterministic dynamical system evolving on a fractal invariant set. Each has had a profound impact on the development of science, and yet their domains of impact remain quite distinct. Despite over a half century of intense research, there is still no consensus on how to combine quantum theory and general relativity theory into a supposed "quantum theory of gravity", nor even a consensus about what such a notion means physically. Moreover, the unpredictability of nonlinear chaotic systems is generally considered quite unrelated to the indeterminism of quantum measurement.

Development of new ideas based on fractal invariant set may lead to some unification of these three revolutions. These ideas evolve around Einstein's great insight that geometry provides the ultimate expression of the laws of physics (Palmer, 2014).

Buchanan (2014) summarizes Palmer's work on fractals and quantum mechanics as follows. Most physicists now take the view that quantum physics is irreducibly non-deterministic and that nature is fundamentally ruled by chance. Palmer now believes that the theory of dynamical chaos and "strange attractors" — the geometrical structures signifying chaos in dissipative systems — might be the key concepts required to tackle a number of fundamental physics issues, among them is building a sensible theory of quantum gravity. Moreover, he suggests that a return to determinism might follow. In dynamical systems theory, the key object in the description of the long-term dynamics of a dissipative system is its attractor — an invariant set that almost all system trajectories approach asymptotically. In a chaotic system, this is a strange attractor with fractal or multifractal geometry. In the famous Lorenz system, for example, the attractor looks crudely like two intersecting surfaces that resemble a butterfly. On closer inspection it turns out to be an infinitely intricate set of nested surfaces — a fractal set of non-integer dimension. Palmer makes a conjecture — but a natural one — that the dynamics of the stuff of our Universe may be similarly described as approaching some invariant attractor. Palmer refers to this as the "invariant set postulate" — that the Universe is evolving causally and deterministically on (or very close to) some measure-zero, fractal invariant set. This would lead to an attractor. The world of deterministic dynamics, he suggests, already has enough weirdness in it to account for everything in quantum physics — but only if we really take deterministic chaos seriously at the universal scale.

### 7.5.2 *Quantum mechanics and string theory*

In recent years, the string field theory has been proposed for a realistic explanation of quantum mechanical laws. Itzhak and Rychkov (2014) have proposed a link between string field theory and quantum mechanics as the basis of all physics. The essential ingredient is the assumption that all matter is made up of strings and that the only possible interaction is joining/splitting as specified in their version of string field theory. Physicists have long sought to unite quantum mechanics and general relativity, and to explain why both work in their respective domains. First proposed in the 1970s, string theory resolved inconsistencies of quantum gravity and suggested that the fundamental unit of matter was a tiny string, not a point, and that the only possible interactions of matter are strings either joining or splitting. At present, no single set of rules can be used to explain all of the physical interactions that occur in the observable universe (Perkins 2014). Hooft (2014) suggests that superstrings may well form exactly the right mathematical system that can explain how quantum mechanics can be linked to a deterministic picture of our world.

### 7.5.3 *Fluid mechanics and quantum mechanics*

Bush (2014) states that thinking of space and time as a liquid might help reconcile quantum mechanics and relativity. If space-time is like a liquid — a concept some physicists say could help resolve a confounding disagreement between two dominant theories (quantum mechanics and relativity) in physics — it must be a very special liquid indeed. A recent study compared astrophysical observations with predictions based on the notion of fluid space-time, and found the idea only works if space-time is incredibly smooth and freely flowing — in other words, a superfluid (Moskowitz, 2014).

### 7.5.4 *General systems theory for fractal space-time fluctuations and quantum-like chaos in atmospheric flows*

Atmospheric flows, a representative example of turbulent fluid flows, exhibit long-range spatiotemporal correlations manifested as the fractal geometry to the global cloud cover pattern concomitant with inverse power-law form for spectra of temporal fluctuations. Such non-local connections are ubiquitous to dynamical systems in nature and are identified as signatures of self-organized criticality (Bak *et al.*, 1988). Applications of self-similarity and self-organized criticality in atmospheric sciences are currently being investigated (Dessai and Walters, 2000; Yano *et al.*, 2012; Craig and Mack, 2013; Stechmann and Neelin, 2014). A cell dynamical system model developed for atmospheric flows shows that the observed long-range spatiotemporal correlations are intrinsic to quantum-like mechanics governing fluid flows (Selvam, 1990, 2007). The model concepts are independent of the exact details such as the chemical, physical, physiological and other properties of the dynamical system and therefore provide a general systems theory applicable to all real world and computed dynamical systems in nature.

The model is based on the concept that spatial integration of enclosed small-scale fluctuations results in the formation of large eddy circulations. The model predicts the following: (a) The flow structure consists of an overall logarithmic spiral trajectory with the quasiperiodic Penrose tiling pattern (Fig. 7.1) for the internal structure such that the ratio of successive eddy radii/circulation speeds is equal to the golden ratio $\tau$ ($\approx 1.618$). (b) Conventional power spectrum analysis will resolve such spiral trajectory as a continuum of eddies with progressive increase in phase (c) Increments in phase angle are concomitant with increase in period length and also represents the variance, a characteristic of quantum systems identified as "Berry's phase". (d) The quantum mechanical constants "fine structure constant" and "ratio of proton mass to electron mass" which are pure numbers and obtained by experimental observations only, are now derived in terms of the golden ratio. (e) Atmospheric flow structure follows Kepler's third law of planetary motion. Therefore, Newton's inverse square law for gravitation applies to eddy masses also. The centripetal accelerations representing inertial masses (of eddies) are equivalent to gravitational masses. Fractal structure to the space-time continuum can be visualized as a nested continuum of vortex (eddy) circulations whose inertial masses obey Newton's inverse square law of gravitation. The model concepts are equivalent to a superstring model for subatomic dynamics, which incorporates gravitational forces. The fractal structure to space-time and also fractalization of microspace is the origin of gravity (Argyris and Ciubotariu, 1997; El Naschie, 2004).

## 7.5.5 *Model predictions and the interpretation of quantum mechanical laws*

The model predictions (Selvam, 1990, 2007) and the interpretation of quantum mechanical laws as applied to macroscale fluid flows are described. It is shown that the apparent paradoxes of quantum mechanics are physically consistent in the context of atmospheric flows.

According to quantum theory, there is no intrinsic upper limit on size or complexity of a physical system above which quantum effects no longer occur (Aspelmeyer and Zeilinger, 2008). Classical physics has equal status with quantum mechanics; it is but a useful approximation of a world that is quantum at all scales. Although quantum effects may be harder to see in the macroworld, the reason has nothing to do with size *per se* but with the way that quantum systems interact with one another. Until the past decade, experimentalists had not confirmed that quantum behavior persists on a macroscopic scale. Today, however, they routinely do. These effects are more pervasive than anyone ever suspected (Vedral, 2011).

### 7.5.5.1 *Probability and amplitude square: Probability of weather system*

Atmospheric flows trace an overall logarithmic spiral trajectory $OR_0R_1R_2R_3R_4R_5$ simultaneously in clockwise and anti-clockwise directions with the quasiperiodic *Penrose* tiling pattern (Steinhardt, 1997) for the internal structure shown in Fig. 7.1.

The spiral flow structure can be visualized as an eddy continuum generated by successive length step growths $OR_0$, $OR_1$, $OR_2$, $OR_3$, ... respectively equal to $R_1$, $R_2$ $R_3$ ... which follow *Fibonacci* mathematical series such that $R_{n+1} = R_n + R_{n-1}$ and $(R_{n+1}/R_n) = \tau$ where $\tau$ is the golden mean equal to $(1+\sqrt{5})/2$ ($\approx 1.618$). Considering a normalized length step equal to 1 for the last stage of eddy growth, the successively decreasing radial length steps can be expressed as 1, $1/\tau$, $1/\tau^2$, $1/\tau^3$, .... The normalized eddy continuum comprises fluctuation length scales 1, $1/\tau$, $1/\tau^2$, .... The probability of occurrence is equal to $1/\tau$ and $1/\tau^2$, respectively, for eddy length scale $1/\tau$ in any one or both rotational (clockwise and anti-clockwise) directions. Eddy fluctuation length of amplitude $1/\tau$, has a probability of occurrence equal to $1/\tau^2$ in both rotational directions, i.e., the square of eddy amplitude represents the probability of occurrence in the eddy continuum. Similar result is observed in the subatomic dynamics of quantum systems which are visualized to consist of the superimposition of eddy fluctuations in wave trains (eddy continuum). Boeyens and Thackeray (2014) discuss the remarkable cosmic occurrence of the golden ratio $\tau$ with reference to space-time, relativity and quantum mechanics. Coldea *et al.* (2010) discovered Golden mean (golden ratio) $\tau$ in quantum world, in the hidden symmetry observed for the first time in solid state matter. Selvam (2013, 2014a) has shown that the amplitude as well as the variance of fractal fluctuations follow the same inverse power law distribution.

### 7.5.5.2 *Non-local connection in weather systems*

*Atmospheric flows*: weather systems in updraft regions and weather dissipation in adjacent down draft regions.

Non-local connections are intrinsic to the space-time geometry of quasiperiodic *Penrose* titling pattern traced by the atmospheric flow pattern visualized as an extended object. The instantaneous non-local connections in the string-like energy flow patterns which represent extended objects such as clouds in atmospheric flows can be visualized as shown in Fig. 7.8. The circulation flow pattern with centre $O$ and radius $OU$ (or $OD$) in Fig. 7.8 represents an eddy. In the medium of propagation, namely, atmosphere (air) in this case, upward motion $U$ represents convection and cloud formation in association simultaneously with cloud dissipation in downward motion $D$. There is an instantaneous non-local connection between the phases of the particles at $U$ and $D$. The same concept can be applied to an extended object (Fig. 7.9) such as a row of clouds represented by the wave function $\psi$ which results from the superimposition of a continuum of eddies. The weather system at Fig. 7.9 is an extended object and the severity of the associated weather at any location is given by the square of eddy amplitude $\psi$ at the location. Since the eddy

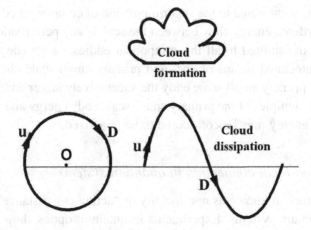

**Figure 7.8:** Instantaneous non-local connection in atmospheric eddy circulations.

## WAVE-PARTICLE DUALITY

### wave-trains in atmospheric flows and cloud formation

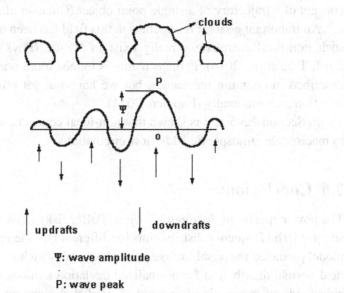

Ψ: wave amplitude

P: wave peak

O: observer

**Figure 7.9:** Wave trains in atmospheric flows and cloud formation.

amplitude $\psi$ is the equal to the superimposition of component eddies with two-way ordered energy flow between the scales, any perturbation at one location is transmitted to all the component eddies. Large eddy circulations are integrated means of enclosed primary small scale circulations. For a fixed primary small-scale eddy the successively larger eddy circulations are in multiples of the primary small-scale eddy energy and therefore large eddy energy may be considered to be quantized.

### 7.5.5.3  *Non-local connection in quantum systems*

The phenomenon known as non-locality or "action at a distance" characterizes quantum systems. Experiments in quantum optics show that two distant events can influence each other instantaneously. Non-local connections in quantum systems apparently violate the fundamental theoretical law in modern physics that signal transmission cannot exceed the speed of light. The distinction between locality and non-locality is related to the concept of a trajectory of a single point object (Chiao *et al.*, 1993).

An important goal for researchers in this field has been to confirm that such non-local correlations really exist in Nature (Phys.org., 20 June 2014; Tura *et al.*, 2014). Perhaps nature is indeed more non-local than is described in quantum mechanics, but we have not yet observed such a situation experimentally (Popescu, 2014).

In Section 7.5.5.2, it is shown that non-local connections are intrinsic to macroscale atmospheric eddy flow circulations.

## 7.6  Conclusions

The power spectra of *Takifugu rubripes* (Puffer fish) DNA base CG density per 10 bp frequency distributions for different DNA lengths follow the model predicted universal inverse power law form which is close to statistical normal distribution for normalized deviation *t* values less than 2 on either side of mean, signifying model predicted quantum-like chaos or long-range correlations in the spatial distribution of CG concentration in the DNA molecule. Such non-local connections enable information communication and control along the total DNA length incorporating coding and non-coding sequences for maintaining robust and optimum performance of vital functions of the living system in a noisy environment. Noncoding DNA sequences are therefore essential for maintenance of functions vital for life. Recent studies show that the proportion of non-coding DNA in the genomic DNA increases with increasing complexity of the organism. The amount of non-coding DNA per genome is a more valid measure of the complexity of an organism than the number of protein-coding genes, and may be related to the emergence of a more sophisticated genomic or regulatory architecture, rather than simply a more sophisticated proteome (Taft and Mattick 2003). One of the steps in turning genetic information into proteins leaves genetic fingerprints, even on regions of the DNA that are not involved in coding for the final protein. They estimate that such fingerprints affect at least a third of the genome, suggesting that while most DNA does not code for proteins, much of it is nonetheless biologically important — important enough, that is, to persist during evolution (Zhang *et al.*, 2008). Much non-coding DNA has a regulatory role (Hayden, 2010; Fraser, 2010).

# Acknowledgement

The author is grateful to Dr. A. S. R. Murty for encouragement during the course of the study.

# References

Andriani, P. and McKelvey, B. (2007). Beyond Gaussian averages: Redirecting management research toward extreme events and power laws, *J. Int. Bus. Stud.*, 38, pp. 1212–1230.

Aparicio, S., Chapman, J., Stupka, E., Putnam, N., Chia, J. M., Dehal, P., Christoffels, A., Rash, S., Hoon, S., Smit, A. *et al.* (2002). Whole-genome shotgun assembly and analysis of the genome of *Fugu rubripes*, *Science*, 297, pp. 1301–1310.

Argyris, J. and Ciubotariu, C. (1997). On El Naschie's complex time and gravitation, *Chaos Solit. Fractals*, 8(5), 743–751.

Aspelmeyer, M. and Zeilinger, A. (2008). Feature: A quantum renaissance, *Physics World*, July, pp. 22–28.

Baake, M. (2002) Quasicrystals: An introduction to structure. In *Physical Properties and Applications*, eds. Suck, J. B. *et al.*, p. 17 (Berlin: Springer).

Bak, P. C., Tang, C., and Wiesenfeld, K. (1988). Self-organized criticality, *Phys. Rev. A*, 38, pp. 364–374.

Bates, A. D. and Maxwell, A. (1993). *DNA Topology* (Oxford University Press, Oxford).

Berry, M. V. (1988). The geometric phase, *Sci. Amer.*, December, pp. 26–32.

Boeyens, J. C. A. and Thackeray, J. F. (2014). Number theory and the unity of science, *S. Afr. J. Sci.*, 110(11/12), pp. 5–6.

Brenner, S., Elgar, G., Sandford, R., Macrae, A., Venkatesh, B., and Aparicio, S. (1993). Characterization of the *Pufferfish* (*Fugu*) genome as a compact model vertebrate genome, *Nature*, 366, pp. 265–268.

Buchanan, M. (2009). Can fractals make sense of the quantum world? *New Scientist*, March (2701), pp. 37–39.

Buchanan, M. (2014). Does not compute? *Nat. Phys.*, 10, p. 404.

Bush, J. W. M. (2015). Pilot-wave hydrodynamics, *Annu. Rev. Fluid Mech.*, 47, pp. 269–292.

Capra, F. (1996). *The Web of Life* (Harper Collins, London).

Check, E. (2006). It's the junk that makes us human, *Nature*, 444, pp. 130–131.

Chiao, R. Y., Kwiat, P. G. and Steinberg, A. M. (1993). Faster than light, *Sci. Am.* Aug., pp. 38–46.

Coldea, R., Tennant, D.A., Wheeler, E.M., Wawrzynska, E., Prabhakaran, D., Telling, M., Habicht, K., Smeibidl, P., and Kiefer, K. (2010). Quantum criticality in an ising chain: Experimental evidence for emergent E8 symmetry, *Science*, 327(5962), pp. 177–180.

Cole, B. J. (2002). Evolution of self-organized systems, *Biol. Bull.*, 202, pp. 256–261.

Craig, G. C., and Mack, J. M. (2013). A coarsening model for self-organization of tropical convection, *J. Geophys. Res. Atmos.*, 118, pp. 8761–8769.

Crollius, H. R., Jaillon, O., Bernot, A., Dasilva, C., Bouneau, L., Fischer, C., Fizames, C., Wincker, P., Brottier, P., Quétier, F., Saurin, W., and Weissenbach, J. (2000). Estimate of human gene number provided by genome-wide analysis using Tetraodon nigroviridis DNA sequence, *Nat. Genet.*, 25, 235–238.

Dessai, S. and Walter, M. E. (2000) Self-organized criticality and the atmospheric sciences: Selected review, New Findings and Future Directions XE Extreme Events: Developing a Research Agenda for the 21st Century, National Center for Atmospheric Research, August, pp. 34–44. http://www.esig.ucar.edu/extremes/papers/walter.PDF.

El Naschie, M. S. (2004). A review of E-infinity theory and the mass spectrum of high energy particle physics, *Chaos Solit/Fractals*, 19, pp. 209–236.

Elgar, G., Clark, M. S., Meek, S., Smith, S., Warner, S., Edwards, Y. J. K., Bouchireb, N., Cottage, A., Yeo, G. S., Umrania, Y., Williams, G., and Brenner, S. (1999). Generation and analysis of 25 Mb of genomic DNA from the *Pufferfish Fugu rubripes* by sequence scanning, *Genome Res.*, 9, pp. 960–971.

Fraser, G. (2010). Inside the complexity labyrinth, *Physics World*, February, pp. 40–41.

Galtier, N., Piganeau, G., Mouchiroud, D., and Duret, L. (2001). GC-Content evolution in mammalian genomes: The biased gene conversion hypothesis, *Genetics*, 159, pp. 907–911.

Hayden, E. C. (2010). "Life is complicated", *Nature*, 464, pp. 664–667.

Itzhak, B. and Rychkov, D. (2014). Background independent string field theory, arXiv:1407.4699v2.

Jean, R. V. (1994). *Phyllotaxis: A Systemic Study in Plant Morphogenesis* (Cambridge University Press, New York).

Jenkinson, A. F. (1977). A Powerful Elementary Method of Spectral Analysis for use with Monthly, Seasonal or Annual Meteorological Time Series, Branch Memorandum No. 57, London: Meteorological Office, pp. 1–23.

Karsenti, E. (2007). Self-organisation processes in living matter, *Interdiscipl. Sci. Rev.*, 32(2), pp. 1–13.

Karsenti, E. (2008). Self-organization in cell biology: A brief history, *Nat. Rev.* (molecular cell biology), 9, pp. 255–262.

Kepler, T. B., Kagan, M. L. and Epstein, I. R. (1991). Geometric phases in dissipative systems, *Chaos*, 1, pp. 455–461.

Kettlewell, J. (2004). Junk throws up precious secret, BBC News, 12 May. http://news.bbc.co.uk.

Kitano, H. (2002). Computational systems biology, *Nature*, 420, pp. 206–210.

Klir, G. J. (1992). Systems science: A guided tour, *J. Biol. Sys.*, 1, pp. 27–58.

Langowski, J. (2006). Polymer chain models of DNA and chromatin, *Eur. Phys. J. E*, 19, pp. 241–249.

Li, W. and Holste, D. (2005). Universal $1/f$ noise, crossovers of scaling exponents, and chromosome-specific patterns of guanine-cytosine content in DNA sequences of the human genome, *Phys. Rev. E.*, 71, pp. 041910–1–9. http://www.nslij-genetics.org/wli/pub/pre05.pdf.

Macia, E. (2006). The role of aperiodic order in science and technology, *Rep. Prog. Phys.*, 69, pp. 397–441.

Milotti, E. (2002). 1/*f* noise: A pedagogical review. http://arxiv.org/abs/physics/0204033.

Moskowitz, C. (2014). If space-time were a superfluid, would it unify physics — or is the theory all wet? *Scientific American*, June 18. http://www.scientificamerican.com/article/superfluid-spacetime-relativity-quantum-physics/.

Nottale, L. and Schneider, J. (1984). Fractals and non-standard analysis, *J. Math. Phys.*, 25, pp. 1296–1300.

Ord, G. N. (1983). Fractal space-time: A geometric analogue of relativistic quantum mechanics, *J. Phys. A*, 16, pp. 1869–1884.

Palmer, T. N. (2005). Quantum reality, complex numbers, and the meteorological butterfly effect, *Bull. Amer. Meteorol. Soc.*, 86, pp. 519–530.

Palmer, T. N. (2009). The Invariant Set Postulate: A new geometric framework for the foundations of quantum theory and the role played by gravity, *Proc. R. Soc. A*, 465, pp. 1–21.

Palmer, T. N., (2014). Gödel and Penrose: New perspectives on determinism and causality in fundamental physics, *Contemp. Phys.*, 55(3), pp. 157–178.

Peacocke, A. R. (1989). *The Physical Chemistry of Biological Organization* (Clarendon Press, Oxford).

Penrose, R. (1965). Gravitational collapse and space-time singularities, *Phys. Rev. Lett.*, 14, pp. 57–59.

Penrose, R. (1989) *The Emperor's New Mind* (Oxford University Press, Oxford), p. 466.

Perkins, R. (2014). String field theory could be the foundation of quantum mechanics. http://phys.org/news/2014-11-field-theory-foundation-quantummechanics.html.

Phys.org. (2014). Experimentally testing non-locality in many-body systems. http://phys.org/news/2014-06-experimentally-nonlocality-many-body.html.

Poland, D. (2004). The phylogeny of persistence in DNA, *Biophys. Chem.*, 112(2–3), pp. 233–244.

Poland, D. (2005). Universal scaling of the C-G distribution of genes, *Biophys. Chem.*, 117(1), pp. 87–95.

Popescu, S. (2014). Non-locality beyond quantum mechanics, *Nat. Phys.*, 10, pp. 264–270.

Prigogine, I. and Stengers, I. (1988). *Order Out of Chaos*, 3rd Ed. (Fontana Paperbacks, London).

Schlosshauer, M., Kofler, J., and Zeilinger, A. (2013a). Snapshot of foundational attitudes toward quantum mechanics. arXiv:1301.1069v1 (quant-ph).

Schlosshauer, M., Kofler, J., and Zeilinger, A. (2013b). The interpretation of quantum mechanics: From disagreement to consensus? *Ann. Phys.*, 525(4), pp. A51–A54.

Schrodinger, E. (1945). *What is Life? The Physical Aspects of the Living Cell* (Cambridge University Press, New York).

Selvam, A. M. (1990). Deterministic chaos, fractals and quantum-like mechanics in atmospheric flows, *Can. J. Phys.*, 68, pp. 831–841. http://xxx.lanl.gov/html/physics/0010046.

Selvam, A. M. (1998). Quasicrystalline pattern formation in fluid substrates and phyllotaxis. In *Symmetry in Plants*, eds. Barabe, D. and Jean, R.V., Vol. 4 (World Scientific, Singapore), pp. 795–809. http://xxx.lanl.gov/abs/chao-dyn/9806001.

Selvam, A. M. (2007). *Chaotic Climate Dynamics* (Luniver Press, United Kingdom).

Selvam, A. M. (2009). Fractal fluctuations and statistical normal distribution, *Fractals*, 17(3), pp. 333–349. http://arxiv.org/pdf/0805.3426.

Selvam, A. M. (2011). Signatures of universal characteristics of fractal fluctuations in global mean monthly temperature anomalies, *J. Sys. Sci. Complexity*, 24, pp. 14–38. http://arxiv.org/pdf/0808.2388.

Selvam, A. M. (2013). Scale-Free Universal spectrum for atmospheric aerosol size distribution for Davos, Mauna Loa and Izana, *Int. J. Bifur. Chaos*, 23(2), pp. 1350028 (13 pages).

Selvam, A. M. (2014a). Universal inverse power law distribution for temperature and rainfall in the UK region, *Dynamics of Atmospheres and Oceans*, 66, pp. 138–150.

Selvam, A. M. (2014b). Universal characteristics of fractal fluctuations in prime number distribution, *Int. J. General Sys.*, 43(8), pp. 828–863.

Sémon, M., Mouchiroud, D., and Duret, L. (2005). Relationship between gene expression and GC-content in mammals: Statistical significance and biological relevance, *Hum. Mol. Genet.*, 14(3), pp. 421–427.

Skipper, M. (2006). Eukaryotic genomes in complete control, *Nat. Rev. Genet.*, 7, pp. 742–743.

Song, C., Havlin, S. and Makse, H. A. (2006). Origins of fractality in the growth of complex networks, *Nat. Phys.*, 2, pp. 275–281.

Spiegel, M. R. (1961). *Statistics* (McGraw-Hill, New York).

Stechmann, S. N. and Neelin, J. D. (2014). First-passage-time prototypes for precipitation statistics, *J. Atmos. Sci.*, 71(9), pp. 3269–3291.

Steinhardt, P. (1997). Crazy crystals, *New Scientist*, 25 January, pp. 32–35.

t Hooft, G. (2014). Superstrings and the foundations of quantum mechanics, *Found Phys.*, 44, pp. 463–471.

Taft R. J. and Mattick, J. S. (2003). Increasing biological complexity is positively correlated with the relative genome-wide expansion of non-protein-coding DNA sequences, *Genome Biol.*, 5, p. 1. http://genomebiology.com/2003/5/1/P1.

Townsend, A. A. (1956). *The Structure of Turbulent Shear Flow*, 2nd Ed. (Cambridge University Press, United Kingdom).

Tura, J., Augusiak, R., Sainz, A. B., Vértesi, T., Lewenstein, M., and Acín, A. (2014). Detecting non-locality in many-body quantum states, *Science*, 344(6189), pp. 1256–1258.

Vedral, V. (2011). Living in a quantum world, *Sci. Am.*, June, pp. 38–43.

Von Bertalanffy, L. (1968). *General Systems Theory: Foundations, Development, Applications* (George Braziller, New York).

Voss, R. F. (1992). Evolution of long-range fractal correlations and $1/f$ noise in DNA base sequences, *Phys. Rev. Lett.*, 68(25), pp. 3805–3808.

Wooley, J. C. and Lin, H. S. (2005). Illustrative problem domains at the interface of computing and biology. In *Catalyzing Inquiry at the Interface of Computing and Biology*, eds. Wooley, J. C. and Lin, H. S. (Computer Science and Telecommunications Board: The National Academies Press, Washington, D.C.), p. 329.

Yano, I., Liu, C. and Moncrieff, M.W. (2012). Self-organized criticality and homeostasis in atmospheric convective organization, *J. Atmos. Sci.*, 69(12), pp. 3449–3462.

Zhang, C., Li, W.-H., Krainer, A. R., and Zhang, M. Q. (2008). An RNA landscape of evolution for optimal exon and intron discrimination, *Proc. Nat. Acad. Sci.*, 105(15), pp. 5797–5802. http://www.pnas.org_cgi_doi_10.1073_pnas.0801692105.

# Chapter 8

# Long-Range Correlations in Human Chromosome X DNA Base CG Frequency Distribution: Data Set VI*

## 8.1 Introduction

Animate and inanimate structures in nature exhibit self-similarity in geometrical shape (Jean, 1994), i.e., parts resemble the whole object in shape. The most fundamental self-similar structure is the forking (bifurcating) structure of tree branches, lung architecture, river tributaries, branched lightning, etc. The hierarchies of complex branching structures represent uniform stretching on a logarithmic scale and belong to the recently identified (Mandelbrot, 1977) category of fractal objects. The complex branching architecture is a self-similar fractal since branching occurs on all scales (sizes) and forms the geometrical shape of the whole object. The substantial identity of the fractal architecture underlying leaf and branch was recognized more than three centuries ago (Arber, 1950). This type of scaling was used by Thompson (1963) in scaling anatomical structures. It appears quite often in the form of *allometric* growth laws in botany as well as in biology (Deering and West, 1992). This particular kind of scaling has been successfully used in biology for over a century. The universal

---

* http://arxiv.org/pdf/physics/0404014.

symmetry of self-similarity underlies apparently irregular complex structures in nature (Schroeder, 1991). The study of fractals now belongs to the new field *nonlinear dynamics and chaos*, a multidisciplinary area of intensive research in all fields of science in recent years (since 1980s) (Gleick, 1987).

Self-similar growth, ubiquitous to nature is therefore governed by universal dynamical laws, which are independent of the exact details (chemical, physical, physiological, electrical, etc.) of the dynamical system. Dynamical systems in nature possess self-similar fractal geometry to the spatial pattern which support fluctuations of processes at all time scales. The temporal fluctuations of dynamical processes are also scale invariant in time and are therefore temporal fractals. The fractal structure to the spatial pattern was pursued as an independent multidisciplinary area of research since late 1970s when the concept of fractals was introduced by Mandelbrot (1977). Though the scale invariant (self-similar) characteristic of temporal fluctuations of dynamical systems has a longer history of observation of more than 30 years (West and Shlesinger, 1989), it was not studied in relation to the associated fractal architecture of spatial pattern till as late as 1988. The unified concept of space-time fractals was identified as a signature of self-organized criticality by Bak *et al.* (1988). For example, the fluctuations in time of atmospheric flows, as recorded by meteorological parameters such as pressure, temperature, wind speed, etc., exhibit self-similar fluctuations in time namely, a zig-zag pattern of increase (decrease) followed by a decrease (increase) on time scales from seconds to years. Such jagged pattern for atmospheric variability (temporal) resembles the self-similar coastline (spatial) structure. Self-similarity indicates long-range correlations, i.e., the amplitude of short- and long-term fluctuations are related to each other by a non-dimensional scale factor alone. Therefore, dynamical laws which govern the space-time fluctuations of smallest scale (turbulence, millimetres —seconds) fluctuations in space-time also apply for the largest scale (planetary, thousands of kilometres — years) in atmospheric flows throughout the globe.

Biological systems exhibit high degree of cooperation in the form of long-range communication. The concept of co-operative existence of fluctuations in the organization of coherent structures has been identified as self-organized co-operative phenomena (Prigogine, 1980) and *synergetics*

(Haken, 1980). The physics of self-organized criticality exhibited by dynamical systems in nature is not yet identified. Selvam (1990) has developed a cell dynamical system model for atmospheric flows which shows that the observed long-range spatiotemporal correlations namely, self-organized criticality are intrinsic to quantum-like mechanics governing turbulent flow dynamics. The model concepts are independent of the exact details, such as the chemical, physical, physiological, etc., properties of the dynamical systems and therefore provide a general systems theory (Peacocke, 1989; Klir, 1992; Jean, 1994; Allegrini *et al.*, 2004) applicable for all dynamical systems in nature.

### 8.1.1 *Model Concepts*

Power spectra of fractal space-time fluctuations of dynamical systems such as fluid flows, stock market price fluctuations, heart beat patterns, etc., exhibit inverse power-law form identified as self-organized criticality (Bak *et al.*, 1988) and represent a self-similar eddy continuum. Li (2004) has given an extensive and informative bibliography of the observed $1/f$ noise, where $f$ is the frequency, in biological, physical, chemical and other dynamical systems. A general systems theory (Selvam, 1990; Selvam and Fadnavis, 1998, 1999) developed by the author shows that such an eddy continuum can be visualized as a hierarchy of successively larger scale eddies enclosing smaller scale eddies. Since the large eddy is the integrated mean of the enclosed smaller eddies, the eddy energy (variance) spectrum follows the statistical normal distribution according to the Central Limit Theorem (Ruhla, 1992). Hence, the additive amplitudes of eddies, when squared, represent the probabilities, which is also an observed feature of the subatomic dynamics of quantum systems such as the electron or photon (Maddox, 1993; Rae, 1998). The long-range correlations intrinsic to self-organized criticality in dynamical systems are signatures of quantum-like chaos associated with the following characteristics: (a) The fractal fluctuations result from an overall logarithmic spiral trajectory with the quasiperiodic Penrose tiling pattern (Selvam, 1990; Selvam and Fadnavis, 1998, 1999) for the internal structure. (b) Conventional continuous periodogram power spectral analyses of such spiral trajectories will reveal a continuum of wavelengths with progressive increase in phase. (c) The broadband power spectrum will have embedded dominant wavebands, the bandwidth increasing with wavelength, and the wavelengths being functions of the golden mean. The first 13 values of the model predicted (Selvam, 1990; Selvam and Fadnavis, 1998, 1999) dominant peak wavelengths are 2.2, 3.6, 5.8, 9.5, 15.3, 24.8, 40.1, 64.9, 105.0, 167.0, 275, 445.0 and 720 in units of the block length 10bp (base pairs) in the present study. Wavelengths (or periodicities) close to the model predicted values have been reported in weather and climate variability (Selvam and Fadnavis, 1998), prime number distribution (Selvam, 2001a), Riemann zeta zeros (non-trivial) distribution (Selvam, 2001b), stock market economics (Sornette *et al.*, 1995; Selvam 2003), DNA base sequence structure (Selvam 2002). (d) The conventional power spectrum plotted as the variance versus the frequency in log–log scale will now represent the eddy probability density on logarithmic scale versus the standard deviation of the eddy fluctuations on linear scale since the logarithm of the eddy wavelength represents the standard deviation, i.e., the root mean square (r.m.s.) value of the eddy fluctuations. The r.m.s. value of the eddy fluctuations can be represented in terms of statistical normal distribution as follows. A normalized standard deviation $t = 0$ corresponds to cumulative percentage probability density equal to 50 for the mean value of the distribution. For the overall logarithmic spiral circulation the logarithm of the wavelength represents the r.m.s. value of eddy fluctuations and the normalized standard deviation $t$ is defined for the eddy energy as

$$t = \frac{\log L}{\log T_{50}} - 1 \tag{8.1}$$

The parameter $L$ in Eq. (8.1) is the wavelength and $T_{50}$ is the wavelength up to which the cumulative percentage contribution to total variance is equal to 50 and $t = 0$. The variable $\log T_{50}$ also represents the mean value for the r.m.s. eddy fluctuations and is consistent with the concept of the mean level represented by r.m.s. eddy fluctuations. Spectra of time series of fluctuations of dynamical systems, for example, meteorological parameters, when plotted as cumulative percentage contribution to total variance versus $t$ follow the model predicted universal spectrum (Selvam, and Fadnavis 1998; see all references under Selvam).

Table 8.1

| s. no. | Accession number | Base pairs from | to | Base pairs used for analysis from | to | Number of data sets | Mean C+G concentration per 10bp | Spectra following normal distribution (%) |
|--------|------------------|------|------|------|---------|------|------|------|
| 1 | NT_078115.2 | 1 | 86563 | 1 | 86520 | 1 | 5.46 | 100 |
| 2 | NT_028413.7 | 1 | 766173 | 1 | 700000 | 10 | 4.79 | 100 |
| 3 | NT_033330.6 | 1 | 623707 | 1 | 560000 | 8 | 4.87 | 100 |
| 4 | NT_025302.12 | 1 | 1381418 | 1 | 1330000 | 19 | 4.38 | 100 |

## 8.2 Data and Analysis

The Human chromosome X DNA base sequence was obtained from the entrez Databases, Homo sapiens Genome (build 34, version 1) at http://www.ncbi.nlm.nih.gov/entrez. The first 4 contiguous data sets (Table 8.1) were chosen for the study. The number of times base C and also base G, i.e., (C+G), occur in successive blocks of 10 bases were determined in successive length sections of 70,000 base pairs giving a series of 7,000 values for each data set giving respectively 10, 8 and 19 data sets for the contiguous data sets 2–4. A data series of 8652 C+G concentration per successive 10 bp values was used for the short contiguous data set 1.

The power spectra of frequency distribution of bases were computed accurately by an elementary, but very powerful method of analysis developed by Jenkinson (1977) which provides a quasicontinuous form of the classical periodogram allowing systematic allocation of the total variance and degrees of freedom of the data series to logarithmically spaced elements of the frequency range (0.5, 0). The cumulative percentage contribution to total variance was computed starting from the high frequency side of the spectrum. The power spectra were plotted as cumulative percentage contribution to total variance versus the normalized standard deviation $t$. The average variance spectra for the Human chromosome X DNA data sets (Table 8.1) and the statistical normal distribution are shown in Fig. 8.1. The "goodness of fit" (statistical chi-square test) between the variance spectra and statistical normal distribution is significant at less than or equal to 5% level for all the data sets (Table 8.1).

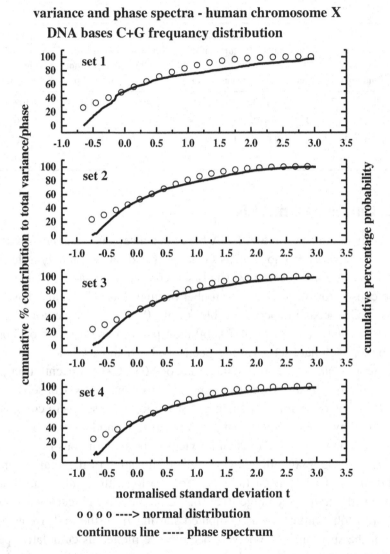

**variance and phase spectra - human chromosome X DNA bases C+G frequancy distribution**

o o o o ----> normal distribution

continuous line ----- phase spectrum

**Figure 8.1:** Average variance spectra for the concentration of base combination C+G per 10 successive base pairs in Human chromosome X DNA. Continuous lines represent the variance spectra and open circles give the statistical normal distribution.

The power spectra exhibit dominant wavebands where the normalized variance is equal to or greater than 1. The dominant peak wavelengths were grouped into 13 class intervals 2–3, 3–4, 4–6, 6–12, 12–20, 20–30, 30–50, 50–80, 80–120, 120–200, 200– 300, 300–600, 600–1000

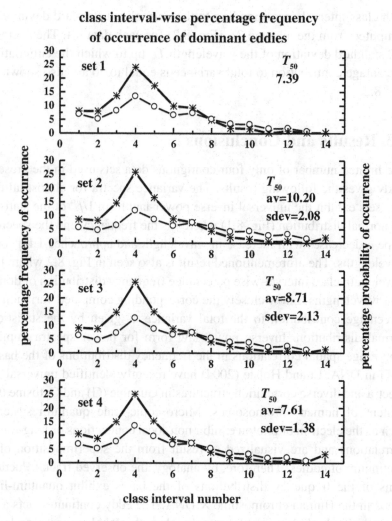

**Figure 8.2:** Average wavelength class interval-wise percentage distribution of dominant (normalized variance greater than 1) wavelengths are given by line + star. The computed percentage contribution to the total variance for each class interval is given by line + open circle. The mean and standard deviation of the wavelengths *t*50 up to which the cumulative percentage contribution to total variance is equal to 50 are also given in the figure.

(in units of 10 bp block lengths) to include the model predicted dominant peak length scales mentioned earlier. Average class interval-wise percentage frequencies of occurrence of dominant wavelengths are shown in Fig. 8.2 along with the percentage contribution to total variance in each class interval corresponding to the normalized standard deviation $t$ computed from the average $T_{50}$ (Fig. 8.2) for each data set. The average and standard deviation of the wavelength $T_{50}$ up to which the cumulative percentage contribution to total variance is equal to 50 are also shown in Fig. 8.2.

## 8.3  Results and Conclusions

The limited number of only four contiguous data sets used in the present study gives the following results. The variance spectra for almost all the data sets exhibit the universal inverse power-law form $1/f^{\alpha}$ of the statistical normal distribution (Fig. 8.1) where $f$ is the frequency and the spectral slope $\alpha$ decreases with increase in wavelength and approaches 1 for long wavelengths. The aforementioned result is also seen in Fig. 8.2 where the wavelength class interval-wise percentage frequency distribution of dominant wavelengths follow closely the corresponding computed variation of percentage contribution to the total variance as given by the statistical normal distribution. Inverse power-law form for power spectra implies long-range spatial correlations in the frequency distributions of the bases C+G in DNA. Li and Holste (2004) have recently identified universal $1/f$ spectra and diverse correlation structures in Guanine (G) and Cytosine (C) content of human chromosomes. Microscopic-scale quantum systems such as the electron or photon exhibit non-local connections or long-range correlations and are visualized to result from the superimposition of a continuum of eddies. Therefore, by analogy, the observed fractal fluctuations of the frequency distributions of the bases exhibit quantum-like chaos in the Human chromosome X DNA. The eddy continuum acts as a robust unified whole fuzzy logic network with global response to local perturbations. Therefore, artificial modification of base sequence structure at any location may have significant noticeable effects on the function of the DNA molecule as a whole. Results of the limited number of data sets in this study indicates that the presence of introns, which do not have meaningful code, may not be redundant, but may serve to organize the effective functioning of the coding exons in the DNA molecule as a complete unit.

# Acknowledgement

The author is grateful to Dr. A. S. R. Murty for encouragement.

# References

Allegrini, P., Giuntoli, M., Grigolini, P., and West, B. J. (2004). From knowledge, knowability and the search for objective randomness to a new vision of complexity, *Chaos Solit. Fractals*, 20, pp. 11–32.

Arber, A. (1950). *The Natural Philosophy of Plant Form* (Cambridge University Press, London).

Bak, P. C., Tang, C., and Wiesenfeld, K. (1988). Self-organized criticality, *Phys. Rev. Ser. A*, 38, pp. 364–374.

Deering, W. and West, B. J. (1992). Fractal Physiology, *IEEE Eng. Med. Biol.*, June, pp. 40–46.

Gleick, J. (1987). *Chaos: Making a New Science* (Viking, New York).

Haken, H. (1980). *Synergetics: An Introduction* (Springer, Berlin).

Jean, R. V. (1994). *Phyllotaxis: A Systemic Study in Plant Morphogenesis* (Cambridge University Press, New York).

Jenkinson, A. F. (1977). A Powerful Elementary Method of Spectral Analysis for use with Monthly, Seasonal or Annual Meteorological Time Series, Meteorological Office, London, Branch Memorandum No. 57, 1977, pp. 1–23.

Klir, G. J. (1992). Systems Science: A Guided Tour, *J. Biol. Sys.*, 1, pp. 27–58.

Li, W. and Holste D. (2005). Universal $1/f$ noise, crossovers of scaling exponents, and chromosome-specific patterns of guanine-cytosine content in DNA sequences of the human genome, *Phys. Rev. E*, 71(4), pp. 041910 (9 pages).

Li, W. (2004). A bibliography on $1/f$ noise. http://www.nslij-genetics.org/wli/1fnoise.

Maddox, J. (1993). Can quantum theory be understood? *Nature*, 361, p. 493.

Mandelbrot, B. B. (1977) *Fractals: Form, Chance and Dimension* (W. H. Freeman, San Francisco).

Peacocke, A. R. (1989). *The Physical Chemistry of Biological Organization* (Clarendon Press, Oxford).

Prigogine, I. (1980). *From Being to Becoming* (Freeman, San Francisco).

Rae, A. (1998). *Quantum-Physics: Illusion or Reality?* (Cambridge University Press, New York).

Ruhla, C. (1992). *The Physics of Chance* (Oxford University Press, Oxford).

Schroeder, M. (1991). *Fractals, Chaos and Power Laws* (W. H. Freeman and Co., New York).

Selvam, A. M. (1990). Deterministic chaos, fractals and quantumlike mechanics in atmospheric flows, *Can. J. Phys.*, 68, pp. 831–841. http://xxx.lanl.gov/html/physics/0010046.

Selvam, A. M. and Fadnavis, S. (1998). Signatures of a universal spectrum for atmospheric interannual variability in some disparate climatic regimes, *Meteorol. Atmos. Phys.*, 66, pp. 87–112. http://xxx.lanl.gov/abs/chaodyn/9805028.

Selvam, A. M. and Fadnavis, S. (1999). Superstrings, cantorian-fractal space-time and quantum-like chaos in atmospheric flows, *Chaos Solit. Fractals*, 10(8), pp. 1321–1334. http://xxx.lanl.gov/abs/chaodyn/9806002.

Selvam, A. M. (2001a). Quantumlike chaos in prime number distribution and in turbulent fluid flows, *APEIRON*, 8(3), pp. 29–64. http://redshift.vif.com/JournalFiles/V08NO3PDF/V08N3SEL.PDF    http://xxx.lanl.gov/html/physics/0005067.

Selvam, A. M. (2001b). Signatures of quantum-like chaos in spacing intervals of non-trivial Riemann zeta zeros and in turbulent fluid flows, *APEIRON*, 8(4), pp. 10–40. http://xxx.lanl.gov/html/physics/0102028, http://redshift.vif.com/JournalFiles/V08NO4PDF/V08N4SEL.PDF.

Selvam, A. M. (2002). Quantum-like chaos in the frequency distributions of the bases A, C, G, T in Drosophila DNA, *APEIRON*, 9(4), pp. 103–148. http://redshift.vif.com/JournalFiles/V09NO4PDF/V09N4sel.pdf.

Selvam, A. M. (2003). Signatures of quantum-like chaos in Dow-Jones index and turbulent fluid flows, *APEIRON*, 10, pp. 1–28. http://arxiv.org/html/physics/0201006,  http://redshift.vif.com/JournalFiles/V10NO4PDF/V10N4SEL.PDF.

Sornette, D. A., Johansen, A., and Bouchaud, J-P. (1995). Stock market crashes, precursors and replicas. http://xxx.lanl.gov/pdf/cond-mat/9510036.

Thompson, D. W. (1963). *On Growth and Form*, 2nd Ed. (Cambridge University Press).

West, B. J. and Shlesinger, M. F. (1989). On the ubiquity of $1/f$ noise, *Int'l. J. Mod. Phys.*, B3(6), pp. 795–819.

# Appendix

## List of Frequently Used Symbols

$v$   frequency

$\alpha$   exponent of inverse power law

$W$   circulation speed (root mean square) of large eddy

$w$   circulation speed (root mean square) of turbulent eddy

$R$   radius of the large eddy

$r$   radius of the turbulent eddy

$T$   time period of large eddy circulation

$t$   time period of turbulent eddy circulation

$k$   fractional volume dilution rate of large eddy by turbulent eddy fluctuations

$z$   eddy length scale ratio equal to $R/r$

$P$   probability density distribution of fractal fluctuations

# Index

Printed in the United States
by Baker & Taylor Publisher Services